Dynamic and Stochastic Approaches to the Environment and Economic Development

Dynamic and Stochastic Approaches to the Environment and Economic Development

Amitrajeet A Batabyal
Rochester Institute of Technology, USA

NEW JERSEY · LONDON · SINGAPORE · BEIJING · SHANGHAI · HONG KONG · TAIPEI · CHENNAI

Published by

World Scientific Publishing Co. Pte. Ltd.
5 Toh Tuck Link, Singapore 596224
USA office: 27 Warren Street, Suite 401-402, Hackensack, NJ 07601
UK office: 57 Shelton Street, Covent Garden, London WC2H 9HE

British Library Cataloguing-in-Publication Data
A catalogue record for this book is available from the British Library.

DYNAMIC AND STOCHASTIC APPROACHES TO THE ENVIRONMENT AND ECONOMIC DEVELOPMENT

Copyright © 2008 by World Scientific Publishing Co. Pte. Ltd.

All rights reserved. This book, or parts thereof, may not be reproduced in any form or by any means, electronic or mechanical, including photocopying, recording or any information storage and retrieval system now known or to be invented, without written permission from the Publisher.

For photocopying of material in this volume, please pay a copying fee through the Copyright Clearance Center, Inc., 222 Rosewood Drive, Danvers, MA 01923, USA. In this case permission to photocopy is not required from the publisher.

ISBN-13 978-981-277-200-8
ISBN-10 981-277-200-6

Typeset by Stallion Press
Email: enquiries@stallionpress.com

Printed in Singapore.

In Memory of

Amar Nath Batabyal, (1937–2002)
Sutapa Batabyal, (1942–1973)
and
Gouranga Chandra Bhattacharya (1933–1998)

Contents

Acknowledgments ix

Part I. Introduction

Chapter 1 Introduction to Dynamic and Stochastic Approaches to the Environment and Economic Development   1

Part II. Agriculture

Chapter 2 Swidden Agriculture in Developing Countries
With Hamid Beladi   39

Chapter 3 Aspects of Land Use in Slash and Burn Agriculture
With Dug Man Lee   57

Chapter 4 A Dynamic and Stochastic Analysis of Fertilizer Use in Swidden Agriculture
With Gregory J. DeAngelo   67

Part III. Renewable Resources

Chapter 5 On Flood Occurrence and the Provision of Safe Drinking Water in Developing Countries   79

Chapter 6 Renewable Resource Management in Developing Countries: How Long Until Crisis?
With Hamid Beladi   87

Chapter 7 A Stackelberg Game Model of Trade in Renewable Resources with Competitive Sellers
With Hamid Beladi   103

Chapter 8 A Differential Game Theoretic Analysis of International Trade in Renewable Resources
With Hamid Beladi   123

Part IV. Environmental Policy

Chapter 9	Environmental Policy in Developing Countries: A Dynamic Analysis	145
Chapter 10	Dynamic Environmental Policy in Developing Countries with a Dual Economy *With Dug Man Lee*	165
Chapter 11	A Dynamic Analysis of Protection and Environmental Policy in a Small Trading Developing Country *With Hamid Beladi*	189
Chapter 12	Aspects of the Theory of Environmental Policy in Developing Countries *With Hamid Beladi*	213
Chapter 13	Public Versus Personal Welfare: An Aspect of Environmental Policymaking in Developing Countries	227
	Index	241

Acknowledgments

Of the thirteen chapters in this book, eleven have appeared previously in different journals. Therefore, I would like to take this opportunity to express my appreciation to various publishers for granting me permission to reprint these previously published chapters. First, Chapter 2 "Swidden Agriculture in Developing Countries" originally appeared in the *Review of Development Economics*, Vol. 8, pp. 255–265, 2004, Chapter 6 "Renewable Resource Management in Developing Countries: How Long Until Crisis?" originally appeared in the *Review of Development Economics*, Vol. 10, pp. 103–112, 2006, Chapter 7 "A Stackelberg Game Model of Trade in Renewable Resources with Competitive Sellers" originally appeared in the *Review of International Economics*, Vol. 14, pp. 136–147, 2006, Chapter 9 "Environmental Policy in Developing Countries: A Dynamic Analysis" originally appeared in the *Review of Development Economics*, Vol. 2, pp. 293–304, 1998, and I thank Blackwell Publishing for their kind permission to reprint these four papers. Second, Chapter 3 "Aspects of Land Use in Slash and Burn Agriculture" originally appeared in *Applied Economics Letters*, Vol. 10, pp. 821–824, 2003, Chapter 5 "On Flood Occurrence and the Provision of Safe Drinking Water in Developing Countries" originally appeared in *Applied Economics Letters*, Vol. 8, pp. 751–754, 2001, and I thank Taylor and Francis http://www.tandf.co.uk/journals for their assistance in reprinting these two papers. Third, Chapter 10 "Dynamic Environmental Policy in Developing Countries With a Dual Economy" originally appeared in the *International Review of Economics and Finance*, Vol. 11, pp. 191–206, 2006, Chapter 11 "A Dynamic Analysis of Protection and Environmental Policy in a Small Trading Developing Country" originally appeared in the *European Journal of Operational Research*, Vol. 143, pp. 197–209, 2002, and I thank

Elsevier for permission to reprint these two papers. Finally, I note that Chapter 4 "A Dynamic and Stochastic Analysis of Fertilizer Use in Swidden Agriculture" initially appeared in *Economics Bulletin*, Vol. 17, pp. 1–10, 2004, that Chapter 12 "Aspects of the Theory of Environmental Policy in Developing Countries" initially appeared in *Discrete Dynamics in Nature and Society*, Vol. 7, pp. 53–58, 2002, and that Chapter 13 "Public Versus Personal Welfare: An Aspect of Environmental Policymaking in Developing Countries" initially appeared in *Economics Bulletin*, Vol. 15, pp. 1–10, 2007.

This book would not have seen the light of day without the help of two groups of individuals. The first group consists of Hamid Beladi, Gregory J. DeAngelo, and Dug Man Lee, and I thank them for allowing me to include our jointly authored papers in this book. The second group consists of Cass Shellman (at RIT) and Ruby Vazquez (at USU), and I thank these two ladies for all manner of assistance with this book. In addition, I thank the Gosnell endowment at RIT for the necessary financial support.

My wife Swapna and my daughter Sanjana both make sacrifices, and, as a result, I am able to pursue my intellectual interests as best as I see fit. Therefore, I thank them both for their love and their support.

Amitrajeet A. Batabyal
Rochester, NY
July 2007

Chapter 1

INTRODUCTION TO DYNAMIC AND STOCHASTIC APPROACHES TO THE ENVIRONMENT AND ECONOMIC DEVELOPMENT

We begin by outlining some of the more salient issues in contemporary research at the interface of the environment and economic development. Next, we note that until very recently, environmental issues have largely been absent in the analyses of economic development. Fortunately, despite the past neglect of this field, there is now a burgeoning literature in this important field. Even so, we point out that there is still a dearth of theoretical research that uses dynamic and stochastic approaches to construct and analyze models of research questions at the intersection of the environment and economic development. Therefore, following this introductory chapter, the 12 chapters of this book show how dynamic and stochastic approaches can be used to effectively model and thereby shed valuable light on a whole host of hitherto largely unstudied research questions concerning the environment and economic development.

1. Preliminaries

The totality of all the ecological systems in the world constitutes a very large part of what we might call our *natural capital stock*. Following Dasgupta (1996), we can also think of this natural capital stock as our *environmental resource base*. Life, as we know it on planet earth, depends fundamentally on this environmental resource base. Even so, until very recently, discussions of the salience of this environmental resource base were largely absent in economic analyses in general and in studies of economic development in particular.[1] This unfortunate state of affairs has gradually begun to change and this change has certainly been accelerated by the publication of the so-called *Brundtland Report* in 1987 (see Brundtland, 1987).

This report introduced, *inter alia*, the notion of sustainable development to the world and it is fair to say that this notion has now become a rallying point for researchers in and practitioners of economic development. Researchers now generally agree that whatever else the notion of sustainability may mean, at the very least, this notion involves the conservation of either a part or all of the earth's natural capital stock.[2] Given the general interest in the notion of sustainability, researchers have sharpened the focus of this notion by asking specifically what it would take for the process of economic development to be sustainable.

Along with this interest in the notion of sustainable development, a parallel development that has taken place stems from the increasingly poor state in which we find the world's fisheries, forests, and rangelands. These so-called *renewable resources*[3] have been the object of rigorous analysis by economists at least since Gordon (1954). Further, since the publication of Gordon's (1954) seminal paper, research by the biologist Garrett Hardin (1968), the economist Herman Daly (1968), and the mathematician Colin Clark (1973; 1976) has increasingly led to the view that what economists call renewable resources and what ecologists more generally call ecological systems are really *jointly determined* ecological-economic systems whose evolution over time is dependent on dynamic and stochastic forces that are partly ecological and partly

---

1. See Dasgupta (1996), Dasgupta and Ehrlich (1996), and Dasgupta and Maler (1997) for a more detailed corroboration of this claim.

2. See Goldin and Winters (1995), Perrings (1996), Farmer and Randall (1997), and Pezzey (1997) for more on the evolution of and the literature concerning the notion of sustainability.

3. Economists commonly distinguish between resources like fisheries and rangelands that have a natural growth rate and hence are regenerative or renewable and resources like minerals such as coal that do not have a natural growth rate — that is meaningful in the context of typical human lifetimes — and hence are non-renewable or exhaustible.

economic in nature. This view has gained broad acceptance in the last three decades and, as such, it is fair to say that today, natural resource management in general is really all about the optimal management of ecological-economic systems.[4] Further, the field of ecological economics itself is now thriving with the launch of a prominent journal published by Elsevier, *Ecological Economics*, that is dedicated to furthering research in this explicitly interdisciplinary field.

The burgeoning of interest in studying sustainable development and the tremendous growth of the field of ecological economics have together greatly influenced contemporary thinking on research questions concerning the environment and economic development. In particular, economists now no longer treat the environmental resource base of developing nations "as an *indefinitely* large and adaptable capital stock" (Dasgupta, 1996, p. 390; emphasis in original). Similarly, ecologists now understand that it would be a big mistake to "regard the human presence as an inessential component of the ecological landscape" (Dasgupta, 1996, p. 390). Finally, we now have a journal published by Cambridge University Press entitled *Environment and Development Economics* that is devoted to the advancement of research on questions at the interface of the environment and economic development.

Despite the presence of a sizeable and now growing literature on the environment and economic development, it is fair to say that there are *very few* theoretical studies of research questions in this field that explicitly incorporate *dynamic* and *stochastic* approaches into their analyses. This state of affairs is both unfortunate and it provides an incomplete and possibly even erroneous perspective on basic issues concerning the environment and economic development. To see why this might be the case, note that economic development is a *process* and, hence, like all processes, this process is best

---

4. See Walters (1986), Holling (1996), Perrings (1996), Dasgupta and Maler (1997), Batabyal (1999), Batabyal and Beladi (1999), and Batabyal and Yoo (2007) for additional details on this point.

studied from a *dynamic* or intertemporal perspective. Second, if economic development in a nation is to be sustainable then it is incumbent upon this nation to conserve at the very least, as we have already noted, some part of its natural capital stock. Now, such conservation will involve the optimal management of ecological-economic systems[5] that are inherently *dynamic* and *stochastic* in nature.

Given the above state of affairs, the central objective of this book is to demonstrate how dynamic and stochastic approaches can be effectively used to construct and analyze theoretical models that shed valuable light on hitherto largely unstudied research questions at the interface of the environment and economic development. The reader should note well that even though we use the traditional language of economics and refer to renewable natural resources, these resources are dynamic and stochastic ecological-economic systems that are jointly determined. Following this introductory chapter, there are 12 essays (chapters) that are grouped into three parts. Parts II and III are concerned with topics that, broadly speaking, would fall within the ambit of natural resource economics as that term is generally understood in economics. In contrast, Part IV of this book is concerned with questions that traditionally would be the object of inquiry in environmental economics.

More specifically, Chapters 2 through 4 in Part II of this book are concerned with a renewable resource of central importance in developing countries and that resource is swidden agriculture.[6] Chapters 5 through 8 comprise Part III of this book and these chapters focus on renewable resources in general and safe drinking water in particular. Chapter 5 studies the provision of safe drinking water and Chapter 6 focuses on renewable resource management; both analyses are conducted in a closed economy setting. Chapters 7 and 8 also focus on renewable resource management issues but now in an open economy setting. Finally, the five chapters comprising Part IV

---

5. Such systems include renewable resources such as fisheries, forests, and rangelands and exhaustible resources such as minerals.

6. Swidden agriculture is also referred to as slash and burn agriculture and as shifting cultivation. Therefore, in this book, we shall use these three terms interchangeably.

of this book analyze the many facets of environmental policy in developing countries. We now proceed to highlight the contributions of the individual essays in this book.

2. The Individual Essays

2.1. *Agriculture*

2.1.1. *The Fallow Period in Swidden Agriculture*

Small farmers in most tropical developing countries practice swidden agriculture. Of the five essential stages in a swidden cycle, the fifth and final stage is the most salient. In this stage, a cleared parcel of forest land (CPFL) is left *fallow* after one or two harvests. If the CPFL is left fallow for a sufficiently long period of time, then nutrients will revert back to the soil and this will permit the swidden cycle to be repeated. Despite the salience of swidden agriculture in tropical developing countries, there is controversy about the merits of this kind of agriculture. On one hand, Dove (1983), Southgate (1990), and Pearce and Warford (1993) have noted that slash and burn agriculture is environmentally destructive because the land-clearing activities of shifting cultivators is directly linked to massive and deleterious deforestation. On the other hand, Peters and Neuenschwander (1988) and Dufour (1990) have claimed that under some circumstances, swidden agriculture based on long fallow periods can be an ecologically and an economically sustainable practice in tropical forests.

The viability of swidden agriculture in the long run depends crucially on the *length* of the fallow period; hence, this period must be chosen optimally. However, beyond recognizing this basic point, researchers have not explained *theoretically* how the length of the fallow period ought to be chosen by a small farmer. In addition, researchers have *not* studied the ways in which the choice of the fallow period length affects the ecology and the economics of the underlying CPFL. Given this state of affairs, Chapter 2 has three goals. First, it constructs a three-state, semi-Markov theoretic model of a CPFL that has been readied for swidden agriculture. Second, it shows for a small farmer how the dynamic and the stochastic properties of this CPFL can be used to derive two objectives that are ecologically meaningful. Finally, using these two objectives, Chapter 2 discusses a probabilistic approach to the determination of the optimal length of the fallow period in swidden agriculture.

A key feature of the semi-Markov model of a CPFL in Chapter 2 is that the length of the fallow period is a *random* variable. In addition, this chapter notes that leaving the CPFL fallow for a specific time period does not guarantee that it will revert back to the ecologically healthiest state 1. Rare and unpredictable environmental events and farmer error in setting the length of the fallow period may result in the CPFL recovering only to the intermediate state 2. This chapter uses the various transition probabilities of the CPFL along with the mean times spent by the CPFL in each of the three states to derive two ecologically meaningful objective functions. The small farmer of this second chapter optimizes these two ecological objective functions subject to specific constraints. We now outline the first of these two optimization problems, and then we comment on the ecological and the economic meaning of the solution to this optimization problem.

The first optimization problem faced by our small farmer involves the minimization of an ecological criterion subject to an economic constraint. The ecological criterion is the *resilience*[7] of the CPFL in the intermediate state 2 and the economic constraint says that the profits from swidden agriculture must not fall below a certain minimum threshold. Solving this optimization problem gives us two

---

7. The response of ecological-economic systems to perturbations is frequently measured by the notion of resilience. Generally speaking, ecologists distinguish between two kinds of resilience, namely, resilience in the sense of C.S. Holling and resilience in the sense of S.L. Pimm. Resilience in the sense of Pimm (1984) is concerned with measuring the rapidity with which a stable ecological-economic system returns to its original state following a perturbation. In contrast, resilience in the sense of Holling refers to "the amount of disturbance that can be sustained [by an ecological-economic system] before a change in system control or structure occurs" (Holling *et al.*, 1995, p. 50). Because Chapter 2 is concerned with the sustainability of swidden agriculture, this chapter focuses on Holling and not Pimm resilience.

first-order necessary conditions for an optimum. Solving these two equations simultaneously gives us the optimal length of the fallow period and the shadow value of the profit constraint. One of these two first-order necessary conditions tells us that in choosing the length of the fallow period optimally, the small farmer will balance ecological and economic considerations. Specifically, the optimal length of the fallow period will be chosen so that the marginal impact of the length of the fallow period on the probability of the CPFL being in the intermediate state 2 is set equal to the product of the shadow value of the profit constraint and the marginal profit from choosing the fallow period length optimally.

If the fallow period length is chosen in this way, then we can be reasonably sure that the CPFL will be healthy in the long run. From an ecological perspective, this means that the resilience of the CPFL in the intermediate state 2 will be low. In economic terms, this means that the CPFL will provide our small farmer with a flow of profits or a flow of consumptive and non-consumptive net benefits in the long run. Although the dynamic and stochastic analysis in Chapter 2 sheds considerable light on the nature of the fallow period length choice problem, this chapter does *not* model the fact that swidden cultivators typically have a choice as far as what kind of crop they would like to grow on their CPFL. In addition, this chapter also does not study the land quality accumulation decision problem faced by shifting cultivators. These two questions are addressed in Chapter 3.

2.1.2. *Crop and Land Quality Accumulation Choices*

Chapter 3 has three goals. First, this chapter constructs a dynamic model of land use by swidden cultivators when these cultivators can choose whether to grow a cash crop or a food/subsistence crop. Second, the chapter examines the land quality accumulation decision problem faced by shifting cultivators. This part of the Chapter 3 analysis provides a second way that can be used to compute the length of the fallow period optimally. In this way, the Chapter 3 analysis complements the earlier analysis in Chapter 2. Finally, Chapter 3 investigates the effects that the optimal land quality accumulation decision has for the relative price of the food crop in particular and for slash and burn agriculture in general.

The focus in this chapter is on an economy in which small farmers (or swidden cultivators) each have a parcel of cleared forest land and they can choose to grow either a cash crop or a food/subsistence crop on this land. The cash crop requires labor L and high quality land A_{hq} for production. In contrast, the food or subsistence crop can be grown with labor L and low quality land A_{lq}. Let w, s_{hq}, and s_{lq} denote the factor rewards to labor, to high quality land, and to low quality land, respectively, and let r denote the interest rate. Deleterious and unpredictable environmental events can make the small farmer's land unfit for cultivation of either the cash crop or the food crop. Chapter 3 accounts for this possibility by supposing that with instantaneous probability q, $q \in (0, 1)$, the cleared land of the swidden cultivators will become unfit for cultivation. With this contingency in mind, the discount rate of the swidden cultivator is effectively $r + q$.

The small farmer can convert low quality land into high quality land by keeping his land fallow for an appropriate length of time. Put differently, this farmer can choose to accumulate land quality by keeping his land fallow. The reader should note two things. First, in the framework of Chapter 3, optimally choosing the length of time during which the cleared land is to be kept fallow is equivalent to optimally accumulating land quality. Second, the purpose of investing in land quality now is to obtain higher profit from the sale of the cash crop later.

The swidden cultivator under study in Chapter 3 maximizes the profit from fallowing land. This profit consists of the earnings from cash crop cultivation less the foregone earnings from food crop cultivation. The solution to this profit-maximization problem gives us the optimal length of the fallow period. The analysis in Chapter 3 shows that as the return to fallowing land increases, the optimal length of the fallow period — or the optimal length of time during which land quality is accumulated — increases. Second, as our swidden cultivator's discount rate rises, it is optimal to reduce the length of time during which this cultivator's land is fallow.

Two additional questions of interest in the context of slash and burn agriculture are also analyzed in Chapter 3. The first such question concerns the discounted lifetime earnings that accrue to the swidden cultivator as a result of his or her decision to accumulate land quality optimally. The analysis here shows that the factor reward of optimally fallowed high quality land exceeds the factor reward of low quality land. The second such question concerns the impact of parametric changes on the factor reward to low quality land. Comparative statics exercises that are carried out in Chapter 3 result in three specific conclusions. Specifically, we learn that, *inter alia*, the factor reward to low quality land falls with increases in the interest rate and the probability of an environmentally disastrous event.

Chapter 3 concludes with a very basic result about economies with slash and burn agriculture. Specifically, it is shown that given an interest rate r, the relative price of the food crop is likely to be higher in economies where there is high demand for keeping land fallow[8] because of a high value of the "return to fallowing land" parameter θ, a high value of the shift variable V, or a low value of the probability q of an environmentally disastrous event. Why? This is because a high θ, a high V, or a low q means that the factor reward for low quality land s_{lq} is high. In turn, because the interest rate r is given, a high s_{lq} can be expected to exert an upward pressure on the relative price of the food crop.

Dickinson (1972), Farnsworth and Golley (1973), and Eckholm (1976) have noted that swidden cultivators can increase the number of harvests on a particular CPFL before this CPFL must be fallowed by applying natural and/or chemical fertilizers. However, beyond recognizing this essential point, researchers have not *theoretically* analyzed the fertilizer use decision problem faced by swidden cultivators. In addition, keeping in mind the dynamic and the stochastic

---

8. Indirectly, this indicates high demand for the cash crop.

setting in which swidden cultivators typically operate, researchers also have *not* studied the conditions under which it is optimal to use fertilizers. These two questions are analyzed in Chapter 4.

2.1.3. Fertilizer Use in Swidden Agriculture

Chapter 4 focuses on a swidden cultivator with a parcel of land that has just been cleared for the planting of a particular crop. This cultivator would like to repeat as many swidden cycles as possible on the CPFL, but in doing this, (s)he must contend with the deterioration in soil fertility on this CPFL. It is possible to extend the useful life of this CPFL by using fertilizers.[9] However, as noted by Dickinson (1972), Eckholm (1976), and others, fertilizers are *costly*, and because swidden cultivators are typically poor, small-scale farmers, they often will not possess the financial resources to purchase fertilizers. There are two policies available to the swidden cultivator under study. The first policy is a *passive* one in which no fertilizer is used and the cultivator relies solely on natural or environmental factors to delay the deterioration in soil fertility. The second policy is an *active* one in which the cultivator uses fertilizers to decelerate the deterioration in soil fertility.

The essential stock variable that is affected by the repeated planting of the crop in question is the stock of soil fertility. Owing to a variety of reasons, the lowering of the stock of soil fertility is generally *probabilistic* and not deterministic. Therefore, the soil fertility stock is assumed to be a stochastic process that can exist in one of many possible states. State 0 is the best possible state of existence for this stock variable. Further, the stock of soil fertility changes state in accordance with a Wiener or Brownian motion process with drift $\delta > 0$.[10] With repeated planting of the crop in question, soil fertility on the CPFL deteriorates, the Brownian motion process changes state, and eventually this process gets to a "breakdown"

---

9. In the rest of this chapter, when we refer to fertilizer use, we are referring to both natural and to chemical fertilizer use.

10. For more on the Wiener process, see Ross (1996, Chapter 8) and Ross (2003, Chapter 10).

state — denoted by f — in which the land must be fallowed.[11] When this is done, mathematically, the Brownian motion process eventually returns to state 0. The cost of the *passive* or no fertilizer use policy is $c(f)$.

The swidden cultivator's *active* policy involves the use of one or more fertilizers. In particular, if the state of the Brownian motion process is b and the active policy is used, then this policy will be successful in improving soil fertility with probability $p(b)$, and it will be unsuccessful with probability $1 - p(b)$. Further, if the active policy is successful in improving soil fertility, then the Brownian motion process being analyzed returns to state 0. In contrast, if this active policy is unsuccessful in improving soil fertility, then the Brownian motion process goes to state f.[12] The cost of attempting to improve soil fertility actively in state b is $c(b)$.

In this probabilistic setting, Chapter 4 determines whether the active or the passive policy minimizes the long-run average cost per time. The analysis in this chapter shows that the long-run average cost of raising soil fertility with the *active* or fertilizer use policy is given by the ratio of the weighted sum of the two cost expressions $c(b)$ and $c(f)$ to the state b, $0 < b < f$, in which this policy is utilized. Similarly, the long-run average cost of the *passive* policy in which the swidden cultivator relies exclusively on natural or environmental factors to improve soil fertility is given by the ratio of the product of the drift parameter of the Brownian motion process δ and the cost of locating and clearing an alternate parcel of forest land $c(f)$ to the fallow state f. What this Chapter 4 analysis makes abundantly clear is that *given* a specific likelihood function $p(b)$, one can always use calculus to minimize the above-mentioned long-run

---

11. An interesting question that emerges in this context is the determination of the optimal length of time during which the CPFL ought to be fallow. This question is addressed in Batabyal and Beladi (2004) and in Chapter 2.

12. We understand that the failure of the fertilizer use policy does not necessarily mean that soil fertility has declined to such an extent that our Wiener process must go to state f. Chapter 4 makes this assumption primarily for reasons of mathematical tractability. Having said this, we recognize that it is possible that the Wiener process will go to some intermediate state e, where e is worse than state b but better than state f.

average cost function. Even so, the more significant point is that the *choice* between the active policy and the passive policy is, to a large extent, dependent on the likelihood function $p(b)$. In particular, for some specifications of this likelihood function, it makes more sense for the swidden cultivator to use the active or fertilizer use policy and for alternate specifications of this same likelihood function, it is less costly to adopt the passive or no fertilizer use policy. This completes our brief discussion of the three chapters that analyze swidden agriculture in developing countries. We now proceed to comment on the contributions of the four chapters that study renewable resource use and management in developing countries in both closed and open economy settings. Providing safe drinking water to flood victims in a closed economy setting is the subject of Chapter 5.

2.2. Renewable Resources

2.2.1. Flood Victims and Safe Drinking Water

National and international development agencies have increasingly begun to embark on a whole host of schemes to provide safe drinking water (SDW) in developing countries.[13] Now, many of the world's developing nations, particularly those in South Asia, are frequently ravaged by floods. Therefore, when a flood occurs, the question of providing SDW to flood victims assumes particular salience. How should a government agency that is interested in distributing SDW to flood victims, go about its task? Further, how might this agency maximize the net social benefit from the provision of SDW? Finally, given that SDW is a particularly scarce commodity in a flood situation, how likely is it that this agency will be unable to meet the stochastic demand for SDW? Although there are a number of empirical and case study based analyses of drinking water problems in developing countries (see Han *et al.*, 1991; Asthana, 1997; Balint, 1999; Reddy, 1999), and even some studies of drinking water

---

13. For more on this, see Munasinghe (1992), Balint (1999), and Kleemeier (2000).

problems in flood situations (see Haque and Zaman, 1993; Emch, 2000), as Chapter 5 points out, there are *no theoretical* studies of the above three questions. Hence, the purpose of Chapter 5 is to show how queuing theory[14] can be used to effectively model and study these three questions concerning the disbursement of SDW in flood-prone developing countries.

Chapter 5 constructs a queuing theoretic model of a place such as West Bengal in India that is frequently ravaged by floods.[15] A government agency is entrusted with the task of providing SDW to flood victims.[16] This agency sets up a relief center and it imports SDW to this center by means of tanker trucks and it "produces" buckets of SDW in accordance with a Poisson process with rate $\lambda > 0$. The analysis in this chapter supposes that once K buckets of water have been produced, the agency under study will have exhausted its available supply of SDW and, hence, it must wait for more water to appear. Flood victims arrive at this relief center in accordance with a Poisson process with rate $\mu > 0$. Each victim is entitled to one bucket of water per day. Upon receipt of this bucket, the victim leaves the relief center. If this victim happens to arrive at the center when K buckets of water have already been disbursed, then (s)he will have to leave the center empty handed. The specific queuing model used to model these events is the $M/M/1$ queue with finite capacity K.[17]

The first task conducted in Chapter 5 is to determine, from the standpoint of the government agency providing SDW to flood victims, the proportion of all flood victims who find the relief center

---

14. See Ross (2003, Chapter 8) for more on queuing theory.

15. West Bengal is frequently ravaged by floods. As reported in *The Economist* (Anonymous, 2000, September 30, p. 6), in year 2000 alone, upwards of 17 million people were adversely affected by floods in this state and in neighboring Bangladesh.

16. Government agencies assigned the task of flood control typically perform many duties, only one of which is the provision of SDW. Chapter 5 focuses on SDW because of the fundamental importance of SDW in sustaining human life and because this chapter wishes to shed light on hitherto unanswered research questions in flood management.

17. For more details on the $M/M/1$ queue with finite capacity, see Batabyal (1996a) and Ross (2003, pp. 480–496).

filled with buckets of SDW. The analysis conducted here shows that this proportion is equal to the stationary probability (P_K) that the relief center has K buckets of SDW. Having computed this probability, the second task that Chapter 5 undertakes is to formulate and solve an optimization problem for the SDW providing government agency. Specifically, this agency chooses the rate μ at which SDW is provided to flood victims to maximize the social net benefit from the provision of SDW. It is shown that optimality requires the government agency to provide SDW so that the marginal social benefit from water provision is equal to the marginal social cost.

The final task that is carried out in Chapter 5 is the computation of the *likelihood* that the government agency under study will be unable to meet the stochastic demand for SDW in a flood situation. The analysis in this part of the chapter shows that the agency in question will be unable to provide SDW to a flood victim only if this victim arrives at the relief center *after* the agency has run out of SDW for that day. Therefore, the likelihood of interest is actually a particular stationary probability and this probability is shown to depend on the parameters of the queue (λ, μ) and on the capacity of the relief center (K) for providing SDW. Further, in the general case, an increase in μ, the rate at which SDW is provided to flood victims, has an ambiguous impact on the proportion of flood victims who come to the agency's relief center and are unable to obtain SDW. This completes our discussion of the Chapter 5 analysis on the provision of SDW to flood victims. We now proceed to discuss renewable resource management in a closed economy setting when there are potential crisis states to contend with.

2.2.2. *Crisis States and Renewable Resource Management*

Chapter 6 notes that people in developing countries are significantly *dependent* on agriculture and on renewable resources, particularly those renewable resources that are found in their local environment. A renewable resource can reasonably be thought to exist in a finite number of states. Some of these states are desirable and others are undesirable. Also, in both these *sets* of states, some states are better than others. Restricting attention to the undesirable set of states, what is important is that some states are likely to be *irreversible*. In these irreversible or *crisis* states,[18] the resource is so degraded that no matter how hard a manager might try, (s)he will be unable to move the resource to any other state. Given that this is the case, one way to look at the task of renewable resource management is to say that a manager's[19] objective is to *maximize* the amount of time a resource spends in the desirable set of states. One can also say that a resource manager's task is to *minimize* the amount of time the resource spends in the undesirable set of states.

Despite the manager's best efforts, (s)he can never be certain that the managed resource will not hit an irreversible state. Given this state of affairs, how should a developing country resource manager proceed? Chapter 6 studies the properties of the following reasonable approach. First, identify the crisis state. Next, put in place a well-designed plan of action. Even with a well-designed plan of action in place, it is still *possible* that a managed resource will hit a crisis state. Consequently, a key question is this: How long until crisis? In other words, the manager would like to know how long it will take for the resource to hit the crisis state. Further, on what does the answer to this question depend? Finally, is the answer history dependent or independent? In other words, does the answer to the how long until crisis question depend on the state in which the manager's plan of action is put in place? Clearly, answers to these questions are vital for successful renewable resource management in developing countries. Yet, there appear to be no previous studies of these questions. Hence, Chapter 6 provides a theoretical analysis of these hitherto unstudied research questions.

---

18. In the rest of our discussion of the contents of Chapter 6, we shall use the terms "irreversible state" and "crisis state" interchangeably.

19. The manager need not be a single individual. In many developing countries, communities collectively manage renewable resources. For more on this, see Wade (1988) and Dasgupta (1996).

An arbitrary renewable resource is modeled as a discrete-time Markov chain with a finite number of *rank ordered* states $0, 1, 2, \ldots, S$. State 0 is the least desirable state and state S is the most desirable state. Further, state 0 is the only crisis state and it is assumed that the resource manager inherits the resource under study in state S. The focus of attention in this chapter is on two management regimes, the so-called lax and the so-called strict management regimes.

The lax management regime is lax because even though the manager inherits the resource in the best possible state S, his or her managerial actions are largely unsuccessful in keeping the resource away from the undesirable states in general and the crisis state in particular. Therefore, given that the resource is in state S, it is just as likely that the resource will next be in state 0 as it is that it will next be in state $S - 3$. Analysis shows that for the lax management regime, when the number of states of the resource approaches infinity, the expected amount of time until the laxly managed resource hits the crisis state is given by the logarithm of the state in which the manager inherits the resource.

The strict management regime is strict in two senses. First, given that the resource is now in state S, the likelihood of hitting the crisis state next is *not* identical to the likelihood of hitting some other (non-crisis) state. Second, the probability of hitting the crisis state next depends on where the resource is initially. In particular, the closer the initial state is to state S, the less likely it is that the resource will hit the crisis state. Analysis shows that as the number of states approaches infinity, the expected amount of time until the strictly managed resource hits the crisis state is given by twice the logarithm of the state in which the manager inherits the resource less the constant 3. Comparing this result with the corresponding result for the lax management regime, it is clear that it generally takes *longer* for the resource to hit the crisis state with the strict management regime. Specifically, Chapter 6 shows that when $S = 1,000,000$, it takes approximately 14 (25) time periods to hit the crisis state with the lax (strict) management regime.

The analysis of the lax and the strict management regimes in Chapter 6 show that the answer to the how long until crisis question depends on the state in which our resource manager inherits the resource under study. Hence, the initial condition *matters*. More generally, the more desirable the state in which the manager inherits the resource, the longer it will take for the resource to hit the crisis state. Chapter 6 looks at renewable resource management in developing countries that are closed economies. In an open economy setting, a number of hitherto irrelevant issues become germane. Chapters 7 and 8 conduct differential game theoretic analyses of some of these issues and it is to these analyses that we now turn.

2.2.3. *Trade in Renewable Resources with Competitive Sellers*

Renewable resources such as fish, timber, ivory, and rhino horns have been traded between countries for quite some time. However, as noted by Clark (1973), Jablonski (1991), and Pimm *et al.* (1995), a great deal of concern has now been expressed about the declining stock levels of most renewable resources. Therefore, the general and hitherto unstudied question that is addressed in Chapter 7 concerns the effect of the stock dependence or the independence of harvesting costs on the efficacy of trade policy. The Stackelberg differential game theoretic analysis conducted in this chapter shows that the *form* of the harvesting cost function has significant implications for the efficacy of trade policy in promoting the conservation of renewable resources.

This chapter first focuses on the efficacy of open loop unit and *ad valorem* tariffs when the harvest cost function is stock *independent*. It is noted that although open loop tariffs are generally dynamically inconsistent (Karp and Newbery, 1993), for this "stock independent" case, the open loop tariffs *are* dynamically consistent. To grasp this point, consider the case in which these tariffs are dynamically inconsistent. In this latter case, at some time $t > 0$ the buyer would want, if he could, to deviate from the tariff trajectory he announced at the beginning of the game and announce a different tariff trajectory. Being forward looking, the representative *competitive* seller in this chapter will anticipate the buyer's desire to change the tariff trajectory he announced initially and hence this tariff will fail to achieve its intended objectives.

The goal of the importing nation is to obliquely encourage the conservation of the renewable resource. Because the tariffs studied in the first part of this chapter are dynamically consistent, they will indirectly achieve their intended conservation goals. Even so, tariffs are not the ideal policy instruments with which to encourage resource conservation. This is because tariffs target imports and they do not do anything directly to encourage conservation of the renewable resource in the exporting nations.

The second half of Chapter 7 analyzes the efficacy of open loop unit and *ad valorem* tariffs when the harvest cost function is stock *dependent*. In this case, it is shown that the optimal open loop unit and *ad valorem* tariffs are dynamically *inconsistent*. Further, the analysis undertaken in this second half demonstrates that attempts to promote renewable resource conservation by means of trade policies are problematic in more ways than one. To see this, note that the analysis in Karp and Newbery (1993) and in Batabyal (1998a) tells us that even when the open loop unit and *ad valorem* tariffs are dynamically inconsistent — and this happens when the harvest cost function is stock dependent — the buyer in the importing nation will prefer to use these inconsistent trade policies rather than follow a dynamically consistent course of action. However, inconsistent policies are not credible and hence the tariff trajectory announced by the buyer at the beginning of the game will not be believed by the sellers and therefore inconsistent policies will typically fail to achieve their resource conservation objectives.

In contrast, when the harvest cost function is stock independent, the optimal open loop tariffs are dynamically consistent and hence believable by the sellers of the resource. Hence, in this case, the buyer's trade policies (tariffs) will indirectly attain their conservation objectives. Although tariffs are an imperfect way of promoting conservation, the analysis in this chapter shows that they may not work as desired. This is because the credibility of the optimal tariffs depends on the form of the harvest cost function and this form is *not* controllable by the buyer. This crucial point has *not* been recognized previously in the literature on international trade in renewable resources.

The stock dependent cost function is more appropriate for endangered renewable resources. Such resources are endangered in part because adequate *domestic* measures have not been taken in the pertinent countries to prevent overexploitation. It is for these endangered resources — where the apposite domestic conservation measures have not been taken — that imperfect supra-national measures such as trade policies are most needed. Unfortunately, the analysis in Chapter 7 tells us that trade policies are likely to be ineffective (because they are not credible) precisely when they are most needed (when the harvest cost function is stock dependent). Does the basic negative message of Chapter 7 change when a monopsonistic buyer in the importing nation engages in resource trade with a monopolistic — and not competitive — seller in the exporting nation in a Stackelberg differential game? This question is analyzed in Chapter 8.

2.2.4. *Trade in Renewable Resources with a Monopolistic Seller*

There are two key *differences* between exhaustible and renewable resources. First, exhaustible resources do not regenerate but renewable resources do. Second, in contrast with exhaustible resources, renewable resources are often an argument in the utility functions of citizens in resource importing nations. Now, the problem of time inconsistency arises when agents with market power make promises that they would subsequently like to break. This problem — see Karp and Newbery (1993) and Groot *et al.* (2003) — has now been fairly well studied in the exhaustible resources literature. As noted in Karp and Newbery (1993, pp. 882–883), a key insight of this literature is that when (i) the future affects the present, (ii) at least one economic actor has market power and is able to influence the future, and (iii) the actor with market power cannot credibly commit herself to future actions, the problem of time inconsistency is salient. These three features are also present in the models analyzed in Chapter 8. In addition, Karp and Newbery (1993, p. 892) note that the problem of time inconsistency is caused by *stock dependent* costs. Therefore, a key question studied in Chapter 8 is whether stock dependent costs alone account for the time inconsistency of optimal policies or whether other factors can also cause time inconsistency.

The analysis in Chapter 8 concentrates on a single buyer who purchases the resource from a *single* seller in a Stackelberg differential game in which the buyer leads. Initially, the focus in this chapter is on the impact of optimal open loop unit and *ad valorem* tariffs when the harvesting cost function is stock independent. In this situation, Chapter 8 obtains three specific results that are contrary to the findings in Chapter 7. First, when the single buyer of a renewable resource faces a monopolistic seller, the buyer's payoff depends on which tariff she uses. Second, the two optimal open loop unit and *ad valorem* tariffs are *not* equivalent. Third, when the buyer uses both tariffs together, she is able to force the monopolistic seller to behave competitively. This means that the harvest rate of the resource is the same whether a competitive seller faces an optimal tariff of either kind or a monopolistic seller faces optimal unit and *ad valorem* tariffs.

As in Chapter 7, because the tariffs studied in the case of the stock independent harvest cost function are time consistent, they will obliquely achieve their intended conservation aims. Even so, Chapter 8 points out that if an importing nation's goal is to encourage conservation of the renewable resource in the exporting country, then tariffs are not the ideal policy tool because tariffs do not do anything directly to promote conservation of the resource in the exporting nation. Therefore, from a resource conservation standpoint, tariffs are blunt policy instruments.

The final contribution of the first part of Chapter 8 is to note that the result that with a stock independent harvest cost function, the optimal tariffs are time consistent, is *not* general. Even with a stock independent harvest cost function, when either the biological growth function of the resource under study is logistic or when the buyer's utility depends on consumption and on the resource stock, the optimal tariffs are time *inconsistent*.

The second part of Chapter 8 examines the efficacy of open loop unit and *ad valorem* tariffs when the harvest cost function is stock dependent. In this case, the optimal unit and *ad valorem* tariffs are time *inconsistent*. Further, the above discussed three specific results obtained in the first part of Chapter 8 hold once again with a stock dependent cost function. In addition, this second part of Chapter 8 corroborates a key Chapter 7 message concerning the problematic nature of efforts to further resource conservation with tariffs. Specifically, we are reminded that when the harvest cost function is stock independent, the optimal tariffs are time consistent and hence believable by the seller. Therefore, in this case, the buyer's tariffs will indirectly attain their resource conservation aims.

Ideally, tariffs ought not to be used to promote renewable resource conservation. This is because tariffs get at the conservation issue indirectly. However, if the seller is unwilling or unable to take measures in his own nation to further resource conservation, then tariffs are one imperfect instrument with which the seller can be encouraged to take the relevant conservation measures. Even so, tariffs may not function as desired. This is because the believability of the optimal tariffs depends on the form of the harvest cost function and this form is *not* controllable by the buyer. One would think that for threatened resources, where the proper domestic conservation measures have not been taken, imperfect supra-national measures such as trade policies might be useful. In this regard, the message from this and the preceding chapter is identical in one fundamental way. Both chapters tell us that tariffs are likely to be unbelievable and hence futile for threatened renewable resource trade where the harvest cost function is generally stock dependent. This brings us to the third and the last part of this book and this part concerns environmental policy in developing countries.

2.3. Environmental Policy

2.3.1. Environmental Policy with Economic Dualism

Much concern has now been expressed about a developing country (DC) government's ability to commit to environmental policy for any reasonable time period. Indeed, some observers have noted that in the face of pressing employment creation needs, DC governments may initiate the process of establishing pollution control policies but their will to continue with such policies is likely to be limited. This employment/environment question in DCs has received very little attention in the extant literature.[20] Hence, Chapter 9 has two aims. First, it analyzes an employment driven dynamic model of environmental policy in a stylized DC. Second, the chapter shows how the DC government's optimal course of action is closely related to its ability to commit to its announced environmental policy.

This chapter uses the specific factors model to study a small, two-sector, trading DC. The two DC sectors consist of a modern, high-wage, environmentally intensive sector in which production causes pollution. The second sector is the traditional, low-wage, environmentally benign sector in which there is no pollution. Workers migrate from the traditional to the modern sector to obtain higher wages. Although workers, as consumers, are adversely affected by pollution, they do not factor this into their migration decisions. Thus, in the absence of governmental policy, migration takes place too quickly and, hence, there is excessive pollution in the nation under study. In this situation, the first-best policy is to tax pollution directly. However, in many DCs the government does not possess the wherewithal to tax pollution directly. Therefore, this chapter assumes that the DC government operates in a second-best environment in which it controls pollution with a production tax.

Chapter 9 studies the DC government's optimal dynamic environmental policy under three assumptions about its ability to commit to a specific course of action. In the first case, the government

---

20. See Lekakis (1991) and Mehmet (1995) for a more detailed corroboration of this claim.

commits to a tax trajectory for an infinite period of time. In the second case, the DC government commits to a tax trajectory for a finite period of time. Unfortunately, in both these cases, the optimal tax policy is dynamically inconsistent.[21] Therefore, forward looking workers will not believe that the government will carry through with its initially announced policy, and hence, this policy will fail to accomplish its aims. In the third case, the government commits to a tax trajectory for an infinitesimal period of time. In the limiting case in which the period of commitment shrinks to zero, the government's tax policy is time consistent.

Section 3 of Chapter 9 depicts the DC government's optimal dynamic environmental policy when it displays infinite commitment. In certain specific scenarios, the DC government moves toward an equilibrium by starting with a large pollution tax. It then gradually lowers this tax to the steady state level. Specifically, the government's open loop tax policy calls for an activist course of action. In other words, in the model of this chapter, it is typically suboptimal to set a zero tax at any point in the program. This notwithstanding, the government's open loop tax policy is time inconsistent. This means that unless there is some mechanism by which the DC government can be bound to its initially announced tax trajectory, this government will fail to achieve its initially announced employment and environmental objectives.[22]

Section 4 of Chapter 9 studies the case in which the DC government commits to its announced environmental policy for a finite time period. In the resulting Markov perfect equilibrium, under certain conditions, an optimal program once again calls for an activist pollution control policy. Further, while the infinite commitment case called for starting with a high tax and then lowering this tax to its steady state value, the limited commitment case calls for

---

21. The reader will recall that dynamic inconsistency was also the focus of much of the analysis in Chapters 7 and 8.

22. The extent to which the government will fail to achieve its objectives depends on the nature and the direction of deviation from the initially announced tax policy.

equalizing the tax at the beginning and at the end of the program. While this limited commitment scenario is plausible, this equilibrium too is time inconsistent. Hence, pollution and employment in sector 2 will not be reduced, and migration from the traditional to the modern sector will not be slowed.

Given this state of affairs, Section 5 of Chapter 9 examines the case in which the DC government commits to a specific tax policy for an infinitesimal period. In this setting, the chapter focuses on the limiting Markov perfect equilibrium in which the government's period of commitment shrinks to zero.[23] Analysis shows that even when the DC government displays no commitment to its tax policy, the welfare loss from being unable to commit is never as great as the welfare gain from reducing pollution. Consequently, the optimal pollution tax is positive. Put differently, the passive aspect (do nothing) of governmental policy is dominated by the activist aspect (control pollution). This is why the limiting pollution tax is positive.

It is possible to rank the three policies studied in this chapter in terms of the government's preference and the policy's ability to achieve its goals. From the standpoint of the DC government's payoff, the most desirable policy is the open loop policy. This policy permits infinite commitment. The second-best policy is the Markov perfect tax policy with a finite period of commitment. The least desirable policy is the limiting Markov perfect tax policy. In contrast, the ranking in terms of goal attainment is exactly the opposite. The limiting Markov perfect tax policy is credible. Hence, this policy will be able to reduce pollution. The other two policies are *not* credible. Therefore, they will fail to achieve the government's environmental goals. This discussion highlights the DC government's dilemma. The policy which results in the highest payoff to the government is the one that is least desirable from the standpoint of goal attainment and social welfare.

---

23. For an alternate approach to the construction of dynamically consistent policies, see Batabyal (1996b; 1996c).

2.3.2. Environmental Policy in the Presence of an Export Subsidy

A number of DCs have protected their export sectors with subsidies to exporters. Therefore, Chapter 10 studies the conduct of dynamic environmental policy in a DC in which the export and the environmentally benign sector is protected with an export subsidy. The specific question that is analyzed in Chapter 10 is the following. What are the properties of optimal environmental policy when a DC government controls pollution by taxing the production of the good manufactured by the polluting and also the import-competing sector, and when this government is unable to commit to the tax policy it announced at the beginning of its tenure in office? It is shown that, in general, the export subsidy has only a slight impact on the DC government's ability to conduct environmental policy effectively.

Chapter 10 uses a specific factors model of the sort used in Chapter 9 to analyze a *small* DC.[24] One sector of this DC is the traditional, low-wage, and non-polluting sector. This traditional sector is also the export sector, and the DC government *protects* this sector by granting a *subsidy* to exporters. For political reasons, this subsidy cannot be repealed. The second sector is the modern, high-wage sector in which production causes pollution. This pollution is *not* transboundary in nature.[25] The modern sector is also the import-competing sector of the DC.[26]

Initially, the DC economy is in disequilibrium. A movement toward equilibrium involves slowing the rate at which workers migrate from the traditional to the modern sector. Chapter 10 studies the DC government's optimal intertemporal environmental policy under three assumptions about its ability to commit to a

---

24. The focus of Chapter 10 on a small DC means that this DC's own policies do not affect world prices. As such, in the rest of our discussion of Chapter 10, we shall not talk about the terms of trade effects of the DC government's environmental policies. The principal focus of Chapter 10 is on the time consistency/inconsistency of alternate pollution control policies. For more on the terms of trade effects of environmental policies, see Batabyal (1993; 1994a; 1994b).

25. For more on the control of one kind of transboundary pollution, see Batabyal (1996d; 1998b) and Xu and Batabyal (2001a; 2001b).

26. The reader may wish to think of the traditional sector as the agricultural sector, and the modern — possibly the infant industry — sector as the steel sector.

particular policy. In the first (second) case, the government commits to its announced tax policy for an infinite (finite) period of time. In both these cases, the government's optimal tax policy is time inconsistent and, hence, implausible. Therefore, a third case is also studied in which the government commits to its tax policy for an infinitesimal period of time. In this case, in the limit, as this period of commitment approaches zero, the government's tax policy is time consistent.

When the DC government displays infinite commitment, the optimal course of action calls for this government to begin with a zero tax. It then raises this tax over the length of the program, and then lowers the tax so that in the stationary state, the pollution tax is, in fact, a subsidy. As explained in this chapter, a tax at time $t = 0$ involves costs but it has no positive policy effect. This is why the optimal initial tax is zero. Further, on account of the positive export subsidy, the steady state level of labor in the two sectors is suboptimal. Hence, to encourage some migration from the traditional to the modern sector in the stationary state, the government grants a subsidy to the producers of the polluting good.

The open loop tax policy with infinite commitment is time inconsistent. Therefore, Chapter 10 next studies a Markov perfect equilibrium in which the DC government displays finite commitment to its announced environmental policy. A salient finding in this case is that the export subsidy now has no effect on the DC government's optimal environmental policy. This suggests that distortions that are *not* in the polluting sector are far less likely to have a detrimental impact on the DC government's ability to conduct environmental policy effectively. As in the infinite commitment case, the DC government's optimal tax policy is not always activist; specifically, when the modern sector's revenue function is separable in its arguments, whether commitment is infinite or finite has *no* bearing on this government's optimal course of action.

The Markov perfect equilibrium in the finite commitment case is time inconsistent. Analysis shows that when the DC government displays infinitesimal commitment to its announced policy, the limiting Markov perfect pollution tax is zero and this tax is independent of the existing export subsidy. This finding has two implications. First, the welfare loss from being unable to commit to environmental policy swamps the welfare gain from reducing pollution. Second, when the DC government revises the pollution tax continually, this government's environmental policy *is* time consistent and hence credible.

The ranking of the three policies studied in this chapter in terms of the government's preference and the policy's ability to achieve its goals is similar to the ranking in Chapter 9. Therefore, despite the presence of a distortion in the traditional sector, the DC government is, once again, in a situation comparable to that in Chapter 9. Specifically, the policy that results in the highest reward for the government is the one that is least likely to lead to the satisfaction of this government's policy goals. It is in this sense that the fear of observers who have worried that in the face of urgent employment creation needs, DC governments are unlikely to be serious about environmental protection, is justified.

2.3.3. *Environmental Policy in the Presence of an Import Tariff*

Chapter 11 continues with and generalizes the previous analysis of dynamic environmental policy in DCs in Chapter 9. In particular, Chapter 11 makes no assumptions about the form of the underlying revenue functions, and it studies the conduct of environmental policy by a small-trading DC in which a tariff protects the import-competing and the polluting sector. The specific question that is addressed in this chapter is the following. What are the properties of optimal dynamic environmental policy when a DC government controls pollution by taxing the production of the good manufactured by the protected sector, and when this government is not necessarily able to commit to the pollution tax policy it announced at the beginning of its tenure in office?

The Chapter 11 model is very similar to the models used in Chapters 9 and 10. Specifically, this chapter uses a dynamic version of the Ricardo–Viner model[27] to study a small-trading and dualistic DC. The polluting sector of the DC is also the import-competing sector and the government uses a positive tariff to protect this sector. One possible interpretation of this sector is that it is the DC's "infant industry."[28] The second sector is the traditional, low-wage, environmentally benign export sector. The political clout of the import-competing sector is such that the government is unable to remove the existing tariff.

The DC economy is initially in disequilibrium because there are distortions (pollution and the tariff) in the DC economy. A move toward equilibrium requires that the production of the polluting good decline over time. Workers have rational expectations. Chapter 11 analyzes the DC government's optimal dynamic environmental policy under three assumptions about its ability to commit to a particular policy. In the first (second) case, the government commits to a tax trajectory for an infinite (finite) period of time. In both these cases, the government's optimal tax policy is time inconsistent and, hence, not credible. Therefore, a third case is also analyzed in which the government commits to its tax policy for an infinitesimal period of time. In the limiting case in which the period of commitment approaches zero, the government's tax policy is time consistent.

When the DC government displays infinite commitment, in an optimal program, the government's pollution tax depends on the existing tariff. First, at the beginning and at the end of the program, the magnitude of the optimal pollution tax is equal to the magnitude of the existing tariff and both are positive. Second, the optimal pollution tax at an *interior* point in the program is generally larger

---

27. This is a standard model in trade theory. In this model, there are two sectors with a factor of production specific to each of these two sectors and a mobile factor of production (typically labor) that can move between these two sectors. For more on this model, see Krugman and Obstfeld (2000, Chapter 3).

28. An "infant industry" is a nascent indigenous industry. Initially, such industries frequently have high costs, and, hence, they find it difficult to compete with other, more established, foreign industries. In turn, this difficulty is often the justification for the protection of "infant industries." For more on these issues, see Krugman and Obstfeld (2000, pp. 255–257).

than the existing positive tariff. Therefore, in an optimal program, the government begins with a positive pollution tax that is equal to the tariff, then raises this tax, and finally lowers this tax so that in the steady state the pollution tax and the tariff are once again equal and positive.

There are two distortions in the DC economy — the import tariff and pollution — and the government has available to it a single policy instrument, namely, the pollution tax. For there to be an improvement in welfare, the number of policy instruments generally ought to equal the number of distortions. This means that the government will *not* be able to use environmental policy to raise welfare unambiguously.

The optimal open loop tax policy with infinite commitment, although activist in nature, is time inconsistent. Hence, Chapter 11 next analyzes a Markov perfect equilibrium in which the DC government displays finite commitment to its proclaimed environmental policy. Analysis shows that a diminution in the length of commitment results in no qualitative change in the government's optimal tax policy. Hence, in general, the discussion in the previous two paragraphs applies to this limited commitment scenario as well. However, there is one difference. In the infinite commitment case, in the steady state there is a single distortion in the economy (the tariff) and the pollution tax raises welfare. However, in the finite commitment case, the government operates in a second-best environment at all points in time.

The Markov perfect equilibrium in the finite commitment case is also time inconsistent. Analysis shows that when the DC government displays infinitesimal commitment to its proclaimed policy, the limiting Markov perfect pollution tax which calls for continuous policy revision is positive, equal to the tariff, and time consistent. This tells us that even when the DC government's period of commitment is infinitesimal, unlike the result in Chapter 10, it is now *not* optimal for the government to set a zero pollution tax. In other words, the activist course of action dominates the "do nothing" or passive course of action.

Three noteworthy outcomes follow from the analysis in Chapter 11. First, this chapter's model is richer than the model in Chapter 9, and it includes the Chapter 9 model as a special case. Second, this Chapter 11 model is able to shed light on the following salient question: What effect does an existing distortion (the import tariff) have on the DC government's optimal dynamic environmental policy? This question cannot be answered using the Chapter 9 model. Third, this chapter's model also answers the following question: What are the properties of optimal dynamic environmental policy in a second-best environment?[29] This question too cannot be answered with the model in Chapter 9.

2.3.4. *Deficits Versus Surpluses and Discretion in Environmental Policy*

In contemporary times, the connections between the environment and economic development have come to dominate academic and public debate in most parts of the world. The analyses in Chapters 9 through 11 tell us that under certain circumstances, employment creation and environmental protection are competing goals. What this means is that although DCs may begin the process of implementing environmental policies, over time, their *commitment* to such policies is likely to wane. The purpose of Chapter 12 is to study two additional aspects of this basic proposition.

In this regard, the nature of dynamic environmental policy in DCs in the presence of a (possibly) binding *financial* or *budget* constraint has been little studied in the extant literature.[30] Therefore, the first part of Chapter 12 examines the following question. When faced with a self-financing or budget constraint, is it optimal for an environmental authority (EA) to alter the trajectory of pollution taxes

---

29. By second-best environment we mean a situation in which (i) the number of distortions exceeds the number of corrective policy instruments available to the DC government, and (ii) the DC government is unable to tax pollution directly. In Chapter 11, there are two distortions (import tariff and pollution) and one policy instrument (pollution tax).

30. For more on the practical effects of budget constraints on the activities of EAs in China and India, see Sinkule and Ortalano (1995, p. 29) and Dwivedi (1997, pp. 124–125).

over time? Or, depending on the actual expenses incurred, does it make more sense for this EA to run deficits/surpluses? Chapters 9 through 11 have demonstrated that a DC government's announced environmental policy is frequently *dynamically inconsistent*. As such, one can ask what nexus there exists between an EA's preferences and credible environmental policy. Specifically, the question analyzed in the second part of Chapter 12 is this. Should an EA make its preferences about the relative benefits of environmental protection versus production of the polluting good public, or should it keep its preferences private?

The analysis in the first part of Chapter 12 focuses on a small, open, infinite horizon DC whose economy is dualistic. One sector is the traditional sector in which there is no pollution. Attention is concentrated primarily on the second and modern sector in which production causes pollution. The EA maximizes the representative consumer's lifetime utility. There are two constraints on the EA's optimization problem. The first is the polluting sector's budget constraint. The second constraint arises from the EA's optimization problem. Specifically, because the subjective time preference factor equals the market discount factor, an Euler equation describing the representative consumer's consumption in any two time periods must be accounted for.[31]

Solving the EA's optimization problem yields four specific results. First, the marginal utility of consumption equals the shadow value of the polluting sector's resources. Second, there is a wedge between the shadow value of the EA's resources and the private value of consumption. Further, this wedge equals the marginal deadweight loss of the pollution tax measured in terms of the representative consumer's utility. Third, like the representative consumer, the EA also finds it optimal to smooth pollution taxes over time. Finally, and most notably, when faced with a self-financing constraint, the EA ought to set a *constant* pollution tax over time. When its expenditures are unusually high, it will be optimal for the EA to run a

---

31. For more on this, see Obstfeld and Rogoff (1996, p. 3).

deficit. Similarly, when its expenditures are unusually low, the EA should run a surplus.

The analysis in the second part of Chapter 12 examines the nexus between the EA's preferences and credible environmental policy. Initially, attention is focused on the so-called discretionary case. In this case, the EA and the polluting sector play a one-shot game among themselves. In the equilibrium of this one-shot game, the optimal level of pollution equals the EA's type. Further, the equilibrium expected level of pollution equals the expected value of the random variable denoting the EA's type. Finally, this chapter computes the expected loss to the EA in the equilibrium of this discretionary one-shot game. These obtained results are then compared with the corresponding results when the EA displays commitment to its environmental policy.

This comparative exercise demonstrates that environmental policy with commitment results in *lower* social losses than does environmental policy with discretion. In other words, society is better off when the EA is committed to environmental policy. A salient implication of the analysis in this part of Chapter 12 is that the EA will actually prefer a system that mandates secrecy about its true preferences regarding the relative benefits of environmental protection versus production of the polluting good. Practically speaking, this means that it is better to have an EA that displays *commitment* to its environmental policy so that the polluting "industries know what to expect [and] how far to go with respect to changing their production processes …" (Dwivedi, 1997, p. 216).

2.3.5. *Personal Versus Public Welfare in Environmental Policymaking*

The analysis in Chapter 12 shows that when faced with a self-financing constraint, it is optimal for the EA to run a deficit/surplus. Second, social losses are lower when this EA keeps its preferences private. Given these findings, Chapter 13 analyzes the nature of the collaboration between an EA and the polluting sector in a DC when the relative weight that this EA places on *public* versus its *own* welfare is unknown. In particular, this chapter first documents the relevance of this issue by discussing actual instances of environmental policymaking in China and India. Next, within the context of the above-stated general issue, this chapter sheds light on three specific questions for any arbitrary time period t.

The first question concerns the determination of the expected and the actual levels of pollution in the DC's polluting sector. The particular object of interest here is an analysis of the equilibrium of the one-shot game between the EA and the polluting sector. To this end, Chapter 13 introduces the *random* variable $\lambda > 0$ which captures the weight that the EA places on public welfare versus its own welfare. To keep the mathematical analysis tractable, it is assumed that the expectation in period $t - 1$ of the value of λ in period t is unity. An examination of the EA's optimization problem shows that when there is uncertainty about an EA's intentions as far as public versus private welfare is concerned, the expected amount of pollution is the *same* as when λ is known to equal unity. However, the *ex post* uncertainty about the type of EA that the polluting sector is confronted with creates *additional variability* in the actual amount of pollution that arises in the polluting sector of the DC under study.

The second question relates to the computation of the average social loss arising in part from the uncertainty about the relative weight that the EA places on public versus its own welfare. The analysis conducted in the second part of Chapter 13 gives rise to three noteworthy results. First, we learn that as the parameter χ, which measures the cost of pollution relative to that of suboptimal output increases, the expected loss to society decreases. Second, as the uncertainty associated with the output supply shock (σ_z^2) goes up, the average loss to society also goes up. Finally, when the uncertainty associated with the EA's weight over public versus its own welfare (σ_λ^2) increases, once again, the expected loss to society also increases. This last result tells us that as far as environmental policymaking is concerned, DCs need to ensure, to the extent possible, that individuals who are placed in positions of authority are public spirited in the discharge of their official duties.

The third and final question involves solving for the optimal value of the parameter, which measures the relative weight the EA places on public versus its own welfare. The analysis in this part of Chapter 13 begins by considering the case in which λ is predictable, and, hence, there is no uncertainty about the relative weight the EA places on public versus its own welfare. In this case of certainty about the EA's type, the relative weight parameter δ is chosen so that it is equal to the positive wedge between the targeted output level of the polluting good and the actual output level of this same good. However, when $\sigma_\lambda^2 \neq 0$ and, hence, λ is unpredictable, the above-discussed choice of δ is not optimal and we have to contend with the fact that there is a tradeoff between reducing *average* pollution by choosing a positive δ and raising the *variance* of pollution because the EA's preferences are stochastic.

3. Conclusions

There is no gainsaying the fact that there is now a sizeable and growing literature on the environment and economic development. Even so, there are very few theoretical studies of research questions in this field that explicitly integrate *dynamic* and *stochastic* approaches into their analyses. As noted in Section 1 of this introductory chapter, this state of affairs is both regrettable and it provides an incomplete and perhaps even inaccurate perspective on fundamental issues concerning the environment and economic development.

Given this objectionable state of affairs, our objective in this book is to demonstrate how dynamic and stochastic approaches can be effectively used to construct and analyze theoretical models that shed valuable light on hitherto largely unstudied questions at the interface of the environment and economic development. To this end, in this introductory chapter, we have highlighted the ways in which the analyses in the 12 individual chapters collectively help accomplish this book's stated objective. The use of dynamic and stochastic approaches to study research questions at the interface of the environment and economic development is still very much in its infancy. Therefore, in the coming years, one may look forward to many interesting developments in theoretical research in this burgeoning new field of inquiry.

References

Anonymous (2000). Politics this Week. *The Economist*, September 30:6.

Asthana, A.N. (1997). Where the Water is Free but the Buckets are Empty: Demand Analysis of Drinking Water in Rural India. *Open Economies Review*, 8:137–149.

Balint, P.J. (1999). Drinking Water and Sanitation in the Developing World: The Miskito Coast of Nicaragua and Honduras as a Case Study. *Journal of Public and International Affairs* 10:99–117.

Batabyal, A.A. (1993). Should Large Developing Countries Pursue Environmental Policy Unilaterally? *Indian Economic Review* 28:191–202.

Batabyal, A.A. (1994a). An Open Economy Model of the Effects of Unilateral Environmental Policy by a Large Developing Country. *Ecological Economics* 10:221–232.

Batabyal, A.A. (1994b). On the Possibility of Attaining Environmental and Trade Objectives Simultaneously. *Environmental and Resource Economics* 4:545–553.

Batabyal, A.A. (1996a). The Queuing Theoretic Approach to Groundwater Management. *Ecological Modelling* 85:219–227.

Batabyal, A.A. (1996b). Consistency and Optimality in a Dynamic Game of Pollution Control I: Competition. *Environmental and Resource Economics* 8:205–220.

Batabyal, A.A. (1996c). Consistency and Optimality in a Dynamic Game of Pollution Control II: Monopoly. *Environmental and Resource Economics* 8:315–330.

Batabyal, A.A. (1996d). Game Models of Environmental Policy in an Open Economy. *Annals of Regional Science* 30:185–200.

Batabyal, A.A. (1998a). Environmental Policy in Developing Countries: A Dynamic Analysis. *Review of Development Economics* 2:293–304.

Batabyal, A.A. (1998b). Games Governments Play: An Analysis of National Environmental Policy in an Open Economy. *Annals of Regional Science* 32:237–251.

Batabyal, A.A. (1999). Contemporary Research in Ecological Economics: Five Outstanding Issues. *International Journal of Ecology and Environmental Sciences* 25:143–154.

Batabyal, A.A. and Beladi, H. (1999). The Stability of Stochastic Systems: The Case of Persistence and Resilience. *Mathematical and Computer Modelling* 30:27–34.

Batabyal, A.A. and Beladi, H. (2004). Swidden Agriculture in Developing Countries. *Review of Development Economics* 8:255–265.

Batabyal, A.A. and Yoo, S.J. (2007). A Probabilistic Approach to Optimal Orchard Management. *Ecological Economics* 60:483–486.

Brundtland, G.H. (1987). *Our Common Future*. The UN World Commission on Environment and Development. Oxford, UK: Oxford University Press.

Clark, C.W. (1973). The Economics of Overexploitation. *Science* 181:630–634.

Clark, C.W. (1976). *Mathematical Bioeconomics*. New York: Wiley.

Daly, H.E. (1968). On Economics as a Life Science. *Journal of Political Economy* 76:392–406.

Dasgupta, P. (1996). The Economics of the Environment. *Environment and Development Economics* 1:387–428.

Dasgupta, P. and Ehrlich, P.H. (1996). Nature's Housekeeping and Human Housekeeping. *Discussion Paper*, Beijer International Institute of Ecological Economics, Stockhohm, Sweden.

Dasgupta, P. and Maler, K.G. (1997). *The Environment and Emerging Development Issues*, Vol. 1. Oxford, UK: Clarendon Press.

Dickinson, J.C. (1972). Alternatives to Monoculture in the Humid Tropics of Latin America. *Professional Geographer* 24:217–222.

Dove, M.R. (1983). Theories of Swidden Agriculture and the Political Economy of Ignorance. *Agroforestry Systems* 1:85–99.

Dufour, D.L. (1990). Use of Tropical Rainforests by Native Amazonians. *BioScience* 40:652–659.

Dwivedi, O.P. (1997). *India's Environmental Policies, Programmes, and Stewardship*. New York: St. Martin's Press.

Eckholm, E.P. (1976). *Losing Ground*. New York: W.W. Norton.

Emch, M. (2000). Relationships Between Flood Control, Kala-azar, and Diarrheal Disease in Bangladesh. *Environment and Planning A* 32:1051–1063.

Farmer, M.C. and Randall, A. (1997). Policies for Sustainability: Lessons from an Overlapping Generations Model. *Land Economics* 73:608–622.

Farnsworth, E.G. and Golley, F.B. (eds.) (1973). *Fragile Ecosystems*. New York: Springer-Verlag.

Goldin, I. and Winters, L.A. (eds.) (1995). *The Economics of Sustainable Development*. Cambridge, UK: Cambridge University Press.

Gordon, H.S. (1954). The Economic Theory of the Common Property Resource: The Fishery. *Journal of Political Economy* 62:124–142.

Groot, F., Withagen, C. and de Zeeuw, A. (2003). Strong Time-Consistency in the Cartel-versus-Fringe Model. *Journal of Economic Dynamics and Control* 28:287–306.

Han, G., Jiang, F. and Yan, J. (1991). 2000AD: Water Environment Problems of China. *International Journal of Social Economics* 18:174–179.

Haque, C.E. and Zaman, M.Q. (1993). Human Responses to Riverine Hazards in Bangladesh: A Proposal for Sustainable Floodplain Development. *World Development* 21:93–107.

Hardin, G. (1968). The Tragedy of the Commons. *Science* 162:1243–1248.

Holling, C.S., Schindler, D.W., Walker, B.W. and Roughgarden, J. (1995). Biodiversity in the Functioning of Ecosystems: An Ecological Synthesis. *In* C. Perrings, K.G. Maler, C. Folke, C.S. Holling and B.O. Jansson (eds.), *Biodiversity Loss*. Cambridge, UK: Cambridge University Press.

Holling, C.S. (1996). Engineering Resilience Versus Ecological Resilience. *In* P.C. Schulze (ed.), *Engineering Within Ecological Constraints*. Washington, DC: National Academy Press.

Jablonski, D. (1991). Extinctions: A Paleontological Perspective. *Science* 253:754–757.

Karp, L. and Newbery, D.M. (1993). Intertemporal Consistency Issues in Depletable Resources. *In* A.V. Kneese and J.L. Sweeney (eds.), *Handbook of Natural Resource and Energy Economics*, Vol. 3. Amsterdam, The Netherlands: Elsevier.

Kleemeier, E. (2000). The Impact of Participation on Sustainability: An Analysis of the Malawi Rural Piped Scheme Program. *World Development* 28:929–944.

Krugman, P.R. and Obstfeld, M. (2000). *International Economics*, 5th edn. Reading, MA: Addison-Wesley.

Lekakis, J.N. (1991). Employment Effects of Environmental Policies in Greece. *Environment and Planning A* 23:1627–1637.

Mehmet, O. (1995). Employment Creation and Green Development Strategy. *Ecological Economics* 15:11–19.

Munasinghe, M. (1992). *Water Supply and Environmental Management: Developing World Applications*. Boulder, CO: Westview Press.

Obstfeld, M. and Rogoff, K. (1996). *Foundations of International Macroeconomics*. Cambridge, MA: MIT Press.

Pearce, D.W. and Warford, J.J. (1993). *World Without End: Economics, Environment, and Sustainable Development.* Oxford, UK: Oxford University Press.

Perrings, C. (1996). Ecological Resilience in the Sustainability of Economic Development. *In* S. Faucheux, D. Pearce and J. Proops (eds.), *Models of Sustainable Development.* Cheltenham, UK: Edward Elgar.

Peters, W.J. and Neuenschwander, L.F. (1988). *Slash and Burn: Farming in the Third World Forest.* Moscow, ID: University of Idaho Press.

Pezzey, J.C.V. (1997). Sustainability Constraints versus "Optimality" versus Intertemporal Concern, and Axioms versus Data. *Land Economics* 73:448–466.

Pimm, S.L. (1984). The Complexity and Stability of Ecosystems. *Nature* 307:321–326.

Pimm, S.L., Russell, G.J. and Gittleman, J.L. (1995). The Future of Biodiversity. *Science* 269:347–350.

Reddy, V.R. (1999). Pricing of Rural Drinking Water: A Study of Willingness and Ability to Pay in Western India. *Journal of Social and Economic Development* 2:101–122.

Ross, S.M. (1996). *Stochastic Processes,* 2nd edn. New York: John Wiley and Sons.

Ross, S.M. (2003). *Introduction to Probability Models,* 8th edn. San Diego, CA: Academic Press.

Sinkule, B.J. and Ortolano, L. (1995). *Implementing Environmental Policy in China.* Westport, CT: Praeger.

Southgate, D. (1990). The Causes of Land Degradation along Spontaneously Expanding Agricultural Frontiers in the Third World. *Land Economics* 66:93–101.

Wade, R. (1988). *Village Republics.* Cambridge, UK: Cambridge University Press.

Walters, C.J. (1986). *Adaptive Management of Renewable Resources.* New York: Macmillan.

Xu, Q. and Batabyal, A.A. (2001a). Price Competition, Pollution, and Environmental Policy in an Open Economy. *Annals of Regional Science* 35:59–79.

Xu, Q. and Batabyal, A.A. (2001b). A Bertrand Model of Trade and Environmental Policy in an Open Economy. *Keio Economic Studies* 38:53–70.

Chapter 2

SWIDDEN AGRICULTURE IN DEVELOPING COUNTRIES

With Hamid Beladi

Small farmers in many tropical developing countries practice swidden agriculture. A key aspect of swidden agriculture is the time period during which the land is left fallow. This chapter uses a new ecological-economic approach to study the fallow period and to determine the optimal length of this period in swidden agriculture. We first construct a theoretical model of a parcel of forest land that has been cleared for swidden agriculture. We then show how the dynamic and the stochastic properties of this cleared land can be used to derive for a small farmer two objective functions that are ecologically meaningful. Finally, using these two objectives, we discuss a probabilistic approach to the determination of the optimal length of the fallow period. In this approach, the focus of the small farmer is on maintaining the ecological and the economic sustainability of swidden agriculture on the cleared parcel of forest land (CPFL).

1. Introduction

Small farmers in most tropical developing countries practice swidden agriculture.[1] There are five essential stages in the swidden cycle.[2] First, large forest trees are cut down by the farmer, the debris is cleared, and the cut growth is burned. The burning of the forest vegetation clears the ground for planting and releases essential

---

1. Swidden agriculture is also known as slash-and-burn agriculture and as shifting cultivation. As such, in the rest of this chapter, we shall use these three terms interchangeably.

2. For more on this and related issues, see Dove (1983), Peters and Neuenschwander (1988), Pearce and Warford (1993), Brown and Pearce (1994), Swinkels *et al.* (1997), and Coomes *et al.* (2000).

nutrients. As the burned vegetation decays, the organic levels in the soil rise and this enhances the soil's fertility. Second, before rains cause soil erosion and before the ash bed can be blown or leached away, planting commences. This typically involves the dropping of seeds into shallow holes made by dibble sticks. Third, with the onset of the rainy season, regular precipitation leads to rapid plant growth. This rapid growth is occasionally accompanied by the simultaneous growth of weeds. These weeds are regularly removed by the farmer to prevent them from taking nutrients away from the crop under cultivation. Fourth, during the harvesting season, the farmer protects the crop from pests and (s)he often uses simple implements such as finger knives to harvest the grain. In the process of harvesting the grain, the farmer retains some of the best seeds for the next planting. Finally, and this is the crucial stage, the cleared parcel of forest land (CPFL) is left fallow after one or two harvests. Within a couple of years, a closed canopy of secondary forest develops. If the CPFL is left fallow for a sufficiently long period of time, then nutrients will revert back to the soil and this will permit the above described swidden cycle to be repeated.

Despite the salience of swidden agriculture in tropical developing countries, there is some controversy about the merits of this kind of agriculture. On one hand, researchers such as Dove (1983), Southgate (1990), and Pearce and Warford (1993) have criticized this kind of agriculture. In particular, these researchers have pointed out that slash-and-burn agriculture is environmentally destructive because the land clearing activities of shifting cultivators is directly linked to massive and deleterious deforestation. On the other hand, a second group of researchers, including Peters and Neuenschwander (1988) and Dufour (1990), have claimed that under some circumstances, swidden agriculture based on long fallow periods can be an ecologically and an economically sustainable practice in tropical forests.

The viability of swidden agriculture in the long run depends crucially on the length of the fallow period; hence, this period must be chosen optimally. There is no dispute on this basic point and there is now a sizable empirical and case study based literature on this point in particular and on the salience of the fallow period in general.[3] However, beyond recognizing this basic point, researchers have not explained theoretically how the length of the fallow period ought to be chosen by a small farmer. In addition to this, researchers have not studied the ways in which the choice of the fallow period length affects the ecology and the economics of the underlying CPFL.

Given this state of affairs, the present chapter has three goals. First, we construct a theoretical model of a parcel of forest land that has been cleared for swidden agriculture. This model accounts for the ecological and the economic aspects of the fallow period length choice problem. Next, we show how the dynamic and the stochastic properties of this CPFL can be used by a small farmer to derive two objectives that are ecologically meaningful. A distinguishing feature of our approach is that these two objectives are very different from the objectives found in traditional economic (and not ecological-economic) analyses of small farmer behavior.[4] Specifically, the two objectives of this chapter are probabilities that are derived from the dynamic and the stochastic properties of the CPFL. Finally, using these two objectives, we discuss a probabilistic approach to the determination of the optimal length of the fallow period in swidden agriculture. In this approach, the focus of a small farmer is on choosing the length of the fallow period so that slash-and-burn agriculture on the CPFL is ecologically and economically viable in the long run.

2. A Model of a CPFL

2.1. *Preliminaries*

Following the recent suggestion in Perrings (1998), we use a semi-Markov[5] model to analyze the fallow period length choice problem

---

3. For a more detailed corroboration of this claim, see Gleave (1996), Hofstad (1997), Silva-Forsberg and Fearnside (1997), Swinkels *et al*. (1997), Coomes *et al*. (2000), Li *et al*. (2000), Ekeleme *et al*. (2000), and Udaeyo *et al*. (2001).

4. For instance, see the analysis in de Janvry *et al*. (1991) and Fafchamps (1992).

5. Standard textbook accounts of semi-Markov processes can be found in Ross (1996, pp. 213–218; 2000, pp. 395–397). Our discussion of semi-Markov processes and the semi-Markov model are based in part on Ross (1996, pp. 213–218; 2000, pp. 395–397).

of this chapter. To this end, consider a stochastic process with states $0, 1, 2, 3, \ldots$, that is now in state $i, i \geq 0$, and that has the following two properties: First, the probability that it will next enter state $j, j \geq 0$, is given by the transition probability P_{ij}. Second, given that the next state entered is j, the time until the transition from state i to state j is a random variable with a general distribution function $G_{ij}(\cdot)$ where $G_{ij}^c(\cdot) = 1 - G_{ij}(\cdot)$. Now let $z(t)$ denote the state of the process at time t. Then $\{z(t) : t \geq 0\}$ is a semi-Markov process.

Put differently, a semi-Markov process is a Markov chain with one appreciable difference. Whereas a Markov chain spends one unit of time in a state before making a transition to some other state, a semi-Markov process stays in a particular state for a random amount of time before making a transition to some other state. Let K_i denote the distribution function of the time that $\{z(t) : t \geq 0\}$ spends in state i before making a transition to some state and let b_i denote the expectation of this time in state i. Finally, let T_{ii} be the time between successive transitions into state i and let b_{ii} be the expectation of T_{ii}. That is, $b_{ii} = E[T_{ii}]$. We are now in a position to discuss the properties of the CPFL, that is, the subject of this chapter.

2.2. *The Dynamic and Stochastic CPFL*

Consider a dynamic and stochastic parcel of forest land that has been cleared for the planting of a certain crop. From an ecological standpoint, at any specific point in time, this CPFL can exist in any one of three possible states. State 1 is the healthiest state of the CPFL. This state corresponds to the condition of the CPFL at the time of planting for and during the first harvest. The forest land has just been cleared, the ash bed contains a rich array of minerals, and soil fertility is high. As such, assuming rains occur as expected, the small farmer can expect his/her crop harvest to be profit making. State 2 is an intermediate state. In this state, the condition of the CPFL in general and its soil cover and quality in particular are lower than in state 1. However, the CPFL is not endangered in either an ecological or an economic sense. The reader should think of state 2 as corresponding to the small farmer's second harvest season on the CPFL under study. State 3 corresponds to the fallow period of the CPFL. In our modeling setup, the ecological condition of the CPFL in this state is delicate. In other words, if our CPFL is not left fallow after two harvest seasons, then the subsequent growing of crops on this land will not be a viable prospect.[6]

Let us now formalize these remarks. As a result of the small farmer's activities (planting, weeding, harvesting) and unpredictable ecological/environmental factors (droughts, lack of plant nutrients, unusually low soil moisture), our CPFL stays in state 1 for a mean length of time b_1 and then makes a transition either to state 2 with transition probability P_{12}, or to state 3 with transition probability P_{13}.[7] When the CPFL is in state 2, once again because of the previously mentioned reasons, this land will stay in state 2 for a mean length of time b_2 and then move to state 3 with transition probability P_{23}. When in state 3, all crop growing, weeding, and harvesting activities are terminated and the CPFL is left fallow. As a result of this fallowing, the CPFL vegetation gradually recovers and, depending on the length of the fallow period, the CPFL may grow back into a secondary forest. It is important to note that the rate of recovery in the fallow state depends not only on the nature of the small farmer's activities in states 1 and 2 but also on unpredictable ecological/environmental factors. What this means for our purpose is that the length of the fallow period is itself a *random* variable. Denote the mean length of the fallow period by b_3. Leaving the CPFL fallow for a specific time period does not guarantee that it will revert

---

6. In our modeling setup, the fallow period occurs after two harvesting seasons. In swidden agriculture, farmers typically cultivate a plot of land "for a *few* seasons, then move onto another plot, abandoning the first to regenerate naturally over a period often measured in decades" (Barrett, 1999, p. 162, emphasis added). Although the precise number of seasons corresponding to few will depend on a number of factors, we are analyzing this three-state model in order to keep the subsequent mathematical analysis tractable. The reader should note that at the cost of greater algebraic clutter, our analysis can be generalized to any finite number of harvesting seasons.

7. Clearly, this latter possibility arises only if the CPFL must be fallowed after a single harvesting season.

back to the ecologically healthiest state 1. Rare and unpredictable environmental events and farmer error in setting the length of the fallow period, either singly or collectively, may result in the CPFL recovering only to the intermediate state 2. To account for these features of the problem, we suppose that after the CPFL has been fallowed for a time period of mean length b_3, it returns either to state 1 with transition probability P_{31}, or to state 2 with transition probability P_{32}. We now use these dynamic and stochastic attributes of this CPFL and derive two objectives that our small farmer might optimize.

2.3. Two Small Farmer Objectives

2.3.1. The First Objective: An Unconditional Probability

To derive the first farmer objective, it will be necessary to compute the stationary probabilities for our three state CPFL. Formally, we are interested in computing $P_i = \lim_{t \to \infty} \text{Prob}\{z(t) = i / z(0) = j\}$ for any state j and for states $i = 1, 2, 3$. In other words, given that our CPFL is in state j at time $t = 0$, we want to compute the limiting probability, as time approaches infinity, that the CPFL will be in state i. To perform this computation, let us denote the limiting probabilities of the embedded Markov chain of our CPFL[8] by $\pi_i, i = 1, 2, 3$. Now it is well-known — see Equation 7.23 in Ross (2000, p. 396) — that these limiting probabilities satisfy

$$\pi_j = \sum_{i=1}^{i=3} \pi_i P_{ij}, \quad \sum_{j=1}^{j=3} \pi_j = 1. \tag{1}$$

Consequently, using the transition probabilities of the CPFL and Equation (1), we can calculate the required limiting probabilities.

---

8. For additional details on the embedded Markov chain of a semi-Markov process, see the references cited in footnote 5.

These are

$$\pi_1 = \frac{P_{31}}{1 + P_{31} + P_{12}P_{31} + P_{32}}, \quad \pi_2 = \frac{P_{12}P_{31} + P_{32}}{1 + P_{31} + P_{12}P_{31} + P_{32}},$$

$$\pi_3 = \frac{1}{1 + P_{31} + P_{12}P_{31} + P_{32}}. \tag{2}$$

To determine the stationary probabilities (the P_i's) of the CPFL, we now use equation 7.24 in Ross (2000, p. 396). This equation tells us that the P_i's satisfy

$$P_i = \frac{\pi_i b_i}{\sum_{j=1}^{j=3} \pi_j b_j}. \tag{3}$$

Equations (2) and (3) together give us the stationary probabilities that we are after. We get

$$P_1 = \frac{P_{31} b_1}{P_{31} b_1 + (P_{12}P_{31} + P_{32})b_2 + b_3}, \tag{4}$$

and

$$P_2 = \frac{(P_{12}P_{31} + P_{32})b_2}{P_{31} b_1 + (P_{12}P_{31} + P_{32})b_2 + b_3},$$

$$P_3 = \frac{b_3}{P_{31} b_1 + (P_{12}P_{31} + P_{32})b_2 + b_3}. \tag{5}$$

In the context of this chapter's ecological-economic approach to determining the optimal length of the fallow period, these stationary probabilities have a distinct ecological meaning. As discussed in Krebs (1985, p. 587), Perrings (1998), Batabyal (1999; 2000), and Batabyal and Beladi (1999), these probabilities measure the asymptotic resilience of the CPFL in each of these three states. Resilience is an ecological stability property and it refers to "the amount of disturbance that can be sustained [by a CPFL] before a change in system control or structure occurs" (Holling et al., 1995, p. 50). This means that we can think of the resilience of a CPFL as a long-run measure of its well-being. Now, if we rank the three states from this well-being perspective, then it should be clear that our CPFL's well-being is highest in state 1 because in this state soil quality is high, minerals are plentiful, and the land is ripe for planting and subsequent harvesting. From a well-being perspective, state 2 is a middle-of-the-road state because minerals, nutrients, and soil quality in general are at an intermediate level. Finally, the CPFL is least well off in state 3 because in this state its ecological condition is delicate. This is also why the CPFL is fallowed in this state.

Recall that the small farmer leaves the CPFL fallow after two harvesting seasons. Further, the mean length of the fallow period is b_3. With these two remarks and the previous paragraph's discussion of the three states in mind, we can now discuss the first objective for our small farmer. Ideally, this farmer would like to choose the expected length of the fallow period b_3 to *maximize* the long-run probability of being in the ecologically most desirable state 1. However, calculus and Equation (4) tell us that P_1 is convex in b_3. As such, it does not make sense to maximize P_1 over b_3. Given this state of affairs, we suppose that our small farmer chooses b_3 to minimize P_2, the long-run probability of being in the intermediate state 2. Further, note that because these long-run probabilities can also be interpreted as the long-run proportion of time that the CPFL is in a particular state, as shown by Ross (1996, pp. 213–218; 2000, pp. 395–397), by minimizing the long-run proportion of time the CPFL spends in state 2, our small farmer is indirectly increasing the long-run proportion of time that the CPFL spends in the ecologically most desirable state 1. Finally, because the long-run probability of being in state 2 is the resilience of the CPFL in state 2 (see the definition in the previous paragraph), we can think of this minimization exercise as one that involves the concurrent minimization of the CPFL's resilience in the intermediate state.

The reader should note that minimizing resilience does not mean that the small farmer is minimizing disturbance or usage of the CPFL. What it does mean is that in the face of shocks from on-going farming activities and ecological/environmental factors (droughts, lack of plant nutrients, unusually low soil moisture), the small farmer is choosing the length of the fallow period to minimize the proportion of time that the CPFL spends in the intermediate state 2. This is the ecological side of the fallow period length choice problem.

In fragile environments of the sort that we are analyzing in this chapter, it is very important to choose the length of the fallow period carefully. It is clear that if the fallow period is too short, then our CPFL will not have had enough time to revert to state 1. However, this does not mean that the manager should err on the side of caution and, *ceteris paribus*, make the fallow period very long. With rising land and population pressures, Southgate (1990) and Barrett (1999) show that if a small farmer chooses a very long fallow period, then (s)he also foregoes the profits from growing and harvesting crops. Put differently, the ecological and the economic aspects of the farmer's choice problem pull in opposite directions. Consequently, in choosing the length of the fallow period, our small farmer will want to account for this ecology versus economics trade-off optimally.

Let us now focus on the economic side of the fallow period length choice problem in greater detail. Note that swidden agriculture gives rise to economic profits to our small farmer where profits equal the difference between total revenues and costs. First consider the revenue aspect. A well-managed CPFL will provide our small farmer with a flow of consumptive (crops that (s)he consumes himself/herself) and monetary (crops that (s)he sells) benefits over time. On the cost side, it is necessary to account for the opportunity cost of the farmer's labor and other costs pertaining to the growing, weeding, and harvesting of crops. Consequently, in determining the length of the fallow period, our small farmer will pay attention to the economic profits from crop growing activities. We suppose that in order to survive, the small farmer cannot let the economic profits associated with swidden agriculture fall below a certain minimum threshold. Denote this threshold by $\tilde{\Pi}$. This tells us that the economic side of the small farmer's choice problem is given by a constraint on the economic profits to this farmer from the practice of swidden agriculture. This constraint is $\Pi(b_3) \geq \tilde{\Pi}$, where $\Pi'(b_3) \leq 0$. In other words, the constraint or profit function is decreasing in the length of the fallow period b_3.

In order to see why this profit function is decreasing in the length of the fallow period, recall that this function measures the economic profit to our farmer from crop growing on the CPFL. Now, when this land is in the fallow state, it is recovering from two harvesting seasons. Consequently, during this time period, the CPFL is not being used for agriculture. This means that from a use perspective, the longer the fallow period, the lower the economic profits. This is why the profit function $\Pi(\cdot)$ is decreasing in the length of the fallow period b_3. This completes the derivation of the first small farmer objective and our discussion of this farmer's minimization problem.

2.3.2. *The Second Objective: A Conditional Probability*

The second farmer objective also involves working with a probability, but now the focus of the small farmer is a little different. As in the previous section on the first objective, once again we shall take a long-run view of the CPFL. In particular, suppose that at time t the CPFL is in the fallow state 3. By choosing the length of the fallow period b_3, the small farmer can affect the state into which the CPFL will next make a transition. Ideally, the farmer would like this next state to be the ecologically healthiest state, i.e., state 1. To this end, if we let $S(t)$ be the state entered at the first transition after time t, then we can determine the long-run conditional probability that the next state visited after t is 1, given that at time t the CPFL is in state 3 and that the mean length of the fallow period is b_3. In other words, ideally, we would like to compute $\lim_{t \to \infty} \text{Prob}\{S(t) = 1/z(t) = 3\}$ and maximize this long-run probability. However, it does not make much sense to maximize this probability, because it can be shown that this probability too is convex in the length of the fallow period b_3. Consequently, we suppose that our small farmer minimizes $\lim_{t \to \infty} \text{Prob}\{S(t) = 2/z(t) = 3\}$ over b_3.

There are two additional things to note here. First, the farmer's principal concern now is to *minimize* the likelihood of going to state 2, given that the CPFL is currently in the fallow state 3. The reader will note that given the long-run proportion of *time* interpretation of these stationary probabilities, once again, our small farmer is attempting to indirectly increase the amount of time spent in the ecologically healthiest state 1. Second, unlike the probability derived in the previous subsection, $\lim_{t\to\infty} \text{Prob}\{S(t) = 2/z(t) = 3\}$ is a *conditional* probability. We now compute this stationary conditional probability. Elementary probability theory tells us that

$$\lim_{t\to\infty} \text{Prob}\{S(t) = 2/z(t) = 3\} = \lim_{t\to\infty} \frac{\text{Prob}\{S(t) = 2, z(t) = 3\}}{\text{Prob}\{z(t) = 3\}}. \quad (6)$$

The joint probability in the numerator of the right-hand side (RHS) of this equation can be simplified with the aid of Theorem 4.8.4 in Ross (1996, p. 217). The probability in the denominator of the RHS of Equation (7) is simply P_3, the stationary probability (see Equation (5)) of finding the CPFL in state 3. With these simplifications, we get

$$\lim_{t\to\infty} \text{Prob}\{S(t) = 2/z(t) = 3\} = P_{32} \cdot \frac{\int_0^\infty G_{32}^c(w)\,dw}{P_3 b_{33}}. \quad (7)$$

Proposition 4.8.1 in Ross (1996, p. 214) can be used to further simplify the denominator on the RHS of Equation (7). This gives

$$\lim_{t\to\infty} \text{Prob}\{S(t) = 2/z(t) = 3\} = P_{32} \cdot \frac{\int_0^\infty G_{32}^c(w)\,dw}{b_3}. \quad (8)$$

The RHS of Equation (8) is the product of two terms. The first term is the probability of making a transition from state 3 to 2. The second term is the ratio of the integral of the tail distribution of the amount of time the CPFL spends in state 3 before making a transition to state 2 to the mean length of the fallow period in state 3. Following Perrings (1998), we can now give an ecological interpretation to this first term. This term is the transient or the short-run resilience of the CPFL in the fallow state 3.

The small farmer's objective now is to choose b_3 to minimize the long-run conditional probability in Equation (8). This is the ecological side of the fallow period length choice problem. The economic side is similar to that in the previous subsection. In order to survive, our farmer cannot let the economic profits from swidden agriculture fall below a certain minimum threshold. Denote this threshold by $\tilde{\Pi}$. Then, as in the previous subsection, the economic side of the farmer's problem is given by a constraint on the minimum acceptable level of profits from swidden agriculture. This constraint is $\Pi(b_3) \geq \tilde{\Pi}$, where $\Pi'(b_3) \leq 0$. This completes the derivation of the second farmer objective and our discussion of the small farmer's optimization problem. We now analyze these optimization problems and then draw inferences for the fallow period length choice problem that is the subject of this chapter.

3. The Small Framer's Choice Problem with Ecological-Economic Criteria

3.1. *Minimizing the Unconditional Probability*

Recall from the discussion in Section 2.3.1 that the first problem faced by our small farmer involves the minimization of an ecological criterion subject to an economic constraint. The ecological criterion is the resilience of the CPFL in the intermediate state 2 and the economic constraint says that the profits from swidden agriculture must not fall below a certain minimum threshold. Formally, our farmer solves

$$\min_{\{b_3 \geq 0\}} \frac{\{P_{12}P_{31} + P_{32}\}b_2}{P_{31}b_1 + (P_{12}P_{31} + P_{32})b_2 + b_3}, \qquad (9)$$

subject to

$$\Pi(b_3) \geq \tilde{\Pi}. \qquad (10)$$

Now, without loss of generality, suppose that the solution to problem (9), (10) yields an interior minimum at which the constraint binds. Then, omitting the complementary slackness conditions, the

Kuhn–Tucker conditions for a minimum are

$$\frac{-(P_{12}P_{31} + P_{32})b_2}{\{P_{31}b_1 + (P_{12}P_{31} + P_{32})b_2 + b_3\}^2} = \rho \Pi'(b_3), \quad (11)$$

where ρ is the multiplier on the profit constraint, and

$$\Pi(b_3) = \tilde{\Pi}. \quad (12)$$

On solving Equations (11) and (12) simultaneously, we get the optimal length of the fallow period (b_3^*) and the shadow value of the profit constraint (ρ^*). The key condition here is Equation (11). This first-order necessary condition tells us that in choosing the length of the fallow period optimally, the small farmer will balance ecological and economic considerations. Specifically, b_3^* will be chosen so that the marginal impact of the length of the fallow period on the probability of the CPFL being in state 2 (the LHS) is set equal to the product of the shadow value of the profit constraint and the marginal profit from choosing the fallow period length optimally (the RHS).

If the fallow period length is chosen in this way, then we can be fairly sure that the CPFL will be healthy in the long run. From an ecological perspective, this means that the resilience of the CPFL in the intermediate state 2 will be low. In economic terms, this means that the CPFL will provide our small farmer with a flow of profits or a flow of consumptive and non-consumptive net benefits in the long run.

3.2. Minimizing the Conditional Probability

Recall from Section 2.3.2 that in this version of the fallow period length choice problem, the small farmer's principal goal is to choose the length of the fallow period so that the amount of time the CPFL spends in state 2 is small, given that this land is currently in the fallow state 3. Formally, the farmer solves

$$\min_{\{b_3 \geq 0\}} P_{32} \cdot \frac{\int_0^\infty G_{32}^c(w)\,dw}{b_3}, \quad (13)$$

subject to Equation (10).

As stated, this minimization problem is unwieldy. Therefore, to demonstrate our approach, we suppose that the amount of time that the CPFL spends in state 3 before making a transition to state 2 is exponentially distributed. Then, integrating the tail distribution function for an exponentially distributed random variable — see Jeffrey (1995, p. 248) — and then evaluating this integral between the lower and the upper limits, we get

$$\int_0^\infty G_{32}^c(w)dw = \frac{1}{\theta}, \tag{14}$$

where $\theta > 0$ is the parameter of the exponential distribution function. Using Equation (14), our small farmer's minimization problem becomes

$$\min_{\{b_3 \geq 0\}} P_{32} \cdot \frac{1}{\theta b_3} \tag{15}$$

subject to Equation (10). Now, as in the previous subsection, suppose that the solution to problems (15)–(10) yields an interior minimum at which the constraint binds. Then, omitting the complementary slackness conditions, the Kuhn–Tucker conditions for a minimum are

$$\frac{-P_{32}}{\theta(b_3)^2} = \rho \Pi'(b_3). \tag{16}$$

Equation (12), and ρ is the multiplier on the profit constraint.

On solving Equations (12) and (16) simultaneously, we obtain the optimal values of the fallow period length (b_3^*) and the shadow value of the profit constraint (ρ^*). The "ecological-economic" optimality condition is Equation (16). This equation tells us that when the small farmer's main concern is to minimize the long-run likelihood of moving to the intermediate state 2 from the fallow state, (s)he will choose the length of the fallow period so that the marginal impact of the fallow period length on the long-run conditional probability of being in state 3 and then making a transition to state 2 (the LHS) is equal to the product of the shadow value of the profit constraint and the marginal profit from choosing the fallow period length optimally (the RHS).

If b_3^* is chosen in this way, then one can be fairly sure that our CPFL will be sustainable in the long run. Once again, it is important to stress that in the context of this chapter, sustainability has a dual meaning. From an ecological standpoint, sustainable means that in the long run, the CPFL will not be resilient in the intermediate state 2. From an economic standpoint, sustainable means that the CPFL will provide our small farmer with a flow of profits in the long run. We now discuss the salience and the policy implications of this chapter's research in the next section.

4. Conclusions

We addressed three issues in this chapter that, to the best of our knowledge, have not been addressed previously in the literature on swidden agriculture in developing countries. First, we used the theory of semi-Markov processes to provide an ecological-economic characterization of a CPFL on which crops are grown and then the land is fallowed. Next, we used the dynamic and the stochastic properties of this CPFL to derive two objectives for a small farmer that are ecologically meaningful. Finally, we used these two objectives to analyze two fallow period length choice problems from an ecological-economic perspective. In this perspective, the focus of our small farmer is on choosing the length of the fallow period so that the CPFL recovers to the ecologically healthiest state 1 and profits from agriculture do not fall below a minimum threshold.

Five specific policy implications follow from this chapter's research. First, unlike many economics papers on the subject of swidden agriculture, our chapter shows that the practice of successful slash-and-burn agriculture involves paying attention to both the ecology and the economics of the CPFL under consideration. Second, we have shown that by optimizing the long-run objective functions of this chapter, a small farmer will also be simultaneously optimizing the resilience of the appropriate states of the CPFL. Third, this chapter has shown how a small farmer might choose the length of the fallow period optimally in an ecological-economic context. Fourth, from a practical perspective, this chapter sheds light on the transition probabilities that will need to be estimated in order to set up the objective functions described in Equations (9) and (15). In addition to this, the minimization exercises presented previously in the unconditional probability and the conditional probability sections provide the small farmer with two different ways of selecting the length of the fallow period optimally. Finally, our research shows that the ecological and the economic aspects of swidden agriculture can be in conflict. This is particularly relevant in developing countries such as Brazil and Indonesia where the practice of swidden agriculture has often been blamed — see Myers (1994) and Rudel and Roper (1997) — for causing tropical deforestation.

The analysis contained in this chapter can be extended in a number of different directions. In what follows, we suggest three avenues for empirical research on the subject of swidden agriculture. First, it would be useful to ascertain whether extant econometric techniques can be used to estimate the transition probabilities of the CPFL under study. Second, given a specific swidden system, knowledge of the amount of time a CPFL spends in a particular state would be helpful in setting up the objective functions discussed in this chapter. Finally, our analysis of the two minimization problems in Section 3 is based on the assumption that the small farmer's profit function is decreasing in the length of the fallow period. Although this seems reasonable from a theoretical standpoint, it would be useful to conduct empirical research to determine whether this assumption is justified. Studies of the fallow period that incorporate these aspects of the problem into the analysis will provide additional insights into the nexuses between the length of the fallow period and the successful practice of swidden agriculture.

References

Barrett, C.B. (1999). Stochastic Food Prices and Slash-and-Burn Agriculture. *Environment and Development Economics* 4:161–176.

Batabyal, A.A. (1999). Species Substitutability, Resilience, and the Optimal Management of Ecological-Economic Systems. *Mathematical and Computer Modelling* 29:35–43.

Batabyal, A.A. (2000). Quantifying the Transient Response of Ecological-Economic Systems to Perturbations. *Environmental Impact Assessment Review* 20:125–133.

Batabyal, A.A. and Beladi, H. (1999). The Stability of Stochastic Systems: The Case of Persistence and Resilience. *Mathematical and Computer Modelling* 30:27–34.

Brown, K. and Pearce, D.W. (eds.) (1994). *The Causes of Tropical Deforestation*. Vancouver, BC: University of British Columbia Press.

Coomes, O.T., Grimard, F. and Burt, G. (2000). Tropical Forests and Shifting Cultivation: Secondary Forest Fallow Dynamics Among Traditional Farmers of the Peruvian Amazon. *Ecological Economics* 32:109–124.

de Janvry, A., Fafchamps, M. and Sadoulet, E. (1991). Peasant Household Behavior with Missing Markets: Some Paradoxes Explained. *Economic Journal* 101:1400–1417.

Dove, M.R. (1983). Theories of Swidden Agriculture and the Political Economy of Ignorance. *Agroforestry Systems* 1:85–99.

Dufour, D.L. (1990). Use of Tropical Rainforests by Native Amazonians. *BioScience* 40:652–659.

Ekeleme, F., Okezie Akobundu, I., Isichei, A.O. and Chikoye, D. (2000). Influence of Fallow Type and Land-Use Intensity on Weed Seed Rain in a Forest/Savanna Transition Zone. *Weed Science* 48:604–612.

Fafchamps, M. (1992). Cash Crop Production, Food Price Volatility, and Rural Market Integration in the Third World. *American Journal of Agricultural Economics* 74:90–99.

Gleave, M.B. (1996). The Length of the Fallow Period in Tropical Fallow Farming Systems: A Discussion with Evidence from Sierra Leone. *Geographical Journal* 1996:14–24.

Holling, C.S., Schindler, D.W., Walker, B. and Roughgarden, J. (1995). Biodiversity in the Functioning of Ecosystems: An Ecological Synthesis. In C. Perrings, K.G. Maler, C. Folke, C.S. Holling and B.O. Jansson (eds.), *Biodiversity Loss*. Cambridge, UK: Cambridge University Press.

Hofstad, O. (1997). Degradation Processes in Tanzanian Woodlands. *Forum for Development Studies* 0:95–115.

Jeffrey, A. (1995). *Handbook of Mathematical Formulas and Integrals*. San Diego, CA: Academic Press.

Krebs, C.J. (1985). *Ecology*, 3rd edition. New York: Harper and Row.

Li, F., Zhao, S. and Geballe, G. (2000). Water Use Patterns and Agronomic Performance for Some Cropping Systems With and Without Fallow Crops in a Semi-Arid Environment of Northwest China. *Agriculture, Ecosystems, and Environment* 79:129–142.

Myers, N. (1994). Tropical Deforestation: Rates and Patterns. *In* K. Brown and D.W. Pearce (eds.), *The Causes of Tropical Deforestation.* Vancouver, BC: University of British Columbia Press.

Pearce, D.W. and Warford, J.J. (1993). *World Without End: Economics, Environment, and Sustainable Development.* Oxford, UK: Oxford University Press.

Perrings, C. (1998). Resilience in the Dynamics of Economy-Environment Systems. *Environmental and Resource Economics* 11:503–520.

Peters, W.J. and Neuenschwander, L.F. (1988). *Slash and Burn: Farming in the Third World Forest.* Moscow, ID: University of Idaho Press.

Ross, S.M. (1996). *Stochastic Processes*, 2nd edition. New York: Wiley.

Ross, S.M. (2000). *Introduction to Probability Models*, 7th edition. San Diego, CA: Academic Press.

Rudel, T. and Roper, J. (1997). The Paths to Rainforest Destruction: Cross-national Patterns of Tropical Deforestation 1975–90. *World Development* 25:53–65.

Silva-Forsberg, M.C. and Fearnside, P.M. (1997). Brazilian Amazonian Caboclo Agriculture: Effect of Fallow Period on Maize Yield. *Forest Ecology and Management* 97:283–291.

Southgate, D. (1990). The Causes of Land Degradation along Spontaneously Expanding Agricultural Frontiers in the Third World. *Land Economics* 66:93–101.

Swinkels, R.A., Franzel, S., Shepherd, K.D., Ohlsson, E.H. and Ndufa, J.K. (1997). The Economics of Short Rotation Improved Fallows: Evidence from Areas of High Population Density in Western Kenya. *Agricultural Systems* 55:99–121.

Udaeyo, N.U., Umoh, G.S. and Ekpe, E.O. (2001). Farming Systems in Southeastern Nigeria: Implications for Sustainable Agricultural Production. *Journal of Sustainable Agriculture* 17:75–89.

Chapter 3

ASPECTS OF LAND USE IN SLASH AND BURN AGRICULTURE

With Dug Man Lee

In this chapter we first construct a theoretical model of land use by swidden cultivators when these cultivators can choose whether to grow a cash crop or a food/subsistence crop. Second, we study the land quality accumulation decision faced by shifting cultivators and, in the process, we show how to compute the optimal length of time during which cleared land is to be left fallow. Finally, we investigate the implications that the optimal land quality accumulation decision has for the relative price of the food crop in particular and slash and burn agriculture in general.

1. Introduction

Slash and burn agriculture is practiced by small farmers in many tropical developing countries.[1] There are five basic stages in the slash and burn cycle.[2] First, forest trees are slashed by farmers, the debris is cleared, and the cut growth is burned. The burning of the forest vegetation clears the ground for planting and releases key nutrients. As the burned vegetation decays, the organic levels in the soil rise and this enhances the soil's fertility. Second, before rains cause soil erosion and before the ash bed can be blown or leached away, planting commences. Third, with the beginning of the rainy

---

1. Slash and burn agriculture is also known as swidden agriculture and as shifting cultivation. Therefore, in the rest of this chapter these three terms are used interchangeably.

2. For additional details on this and related issues, see Dove (1983), Peters and Nuenschwander (1988), Pearce and Warford (1993), Brown and Pearce (1994), Swinkels *et al.* (1997), Coomes *et al.* (2000), and Batabyal and Beladi (2004).

season, normal precipitation results in rapid plant growth. Fourth, during the harvesting season, farmers protect the crop from pests and they frequently use simple implements such as finger knives to harvest the grain. Finally, the cleared parcel of forest land is left fallow. Within a couple of years, land quality improves and a closed canopy of secondary forest develops. If the cleared parcel of forest land is left fallow for a sufficiently long period of time, i.e., if the swidden cultivators accumulate land quality for a sufficiently long period of time, then nutrients will revert back to the soil and this will permit the above described slash and burn cycle to be repeated.

Researchers generally agree that slash and burn agriculture is an important and widespread phenomenon in many tropical developing countries. The disagreement among researchers concerns the desirability of this kind of agriculture. On the one hand, researchers such as Dove (1983), Southgate (1990), and Pearce and Warford (1993) have argued that slash and burn agriculture is environmentally destructive because the land clearing activities of swidden cultivators is responsible for extensive and detrimental tropical deforestation. On the other hand, a second group of researchers including Peters and Neuenschwander (1988) and Dufour (1990) have asserted that under some circumstances, slash and burn agriculture based on long fallow periods can be an ecologically and an economically sustainable practice in tropical forests.

The viability of slash and burn agriculture in the long run depends critically on the land use decisions made by small farmers. Although there now exists a fairly large empirical and case study based literature on slash and burn agriculture — see Gleave (1996), Hofstad (1997), Silva-Forsberg and Fearnside (1997), Swinkels *et al.* (1997), Coomes *et al.* (2000), Li *et al.* (2000), Ekeleme *et al.* (2000), and Udaeyo *et al.* (2001) — there are virtually no *theoretical* studies of the land use decisions of swidden cultivators.

The theoretical paper that is closest in spirit to our chapter is Batabyal and Beladi (2004). In this paper, Batabyal and Beladi construct a dynamic and stochastic model and they then use this model to set up and solve problems that tell us, *inter alia*, how the fallow period in slash and burn agriculture is to be optimally selected. Even though this is a useful paper, it does not model the fact that swidden cultivators typically have a choice as far as what kind of crop they would like to grow. In addition, the Batabyal and Beladi (2004) paper also does not study the land quality accumulation decision faced by shifting cultivators.

Given this state of affairs, our chapter has three goals. First, we construct a theoretical model of land use by swidden cultivators when these cultivators can choose whether to grow a cash crop or a food/subsistence crop. Second, we study the land quality accumulation decision faced by shifting cultivators and, in the process, we complement the Batabyal and Beladi (2004) analysis by showing how to compute the optimal length of time during which the cleared land is to be left fallow. Finally, we investigate the implications that the optimal land quality accumulation decision has for the relative price of the food crop in particular and for slash and burn agriculture in general.

The rest of this chapter is organized as follows. Section 2 first presents a dynamic and stochastic framework and then it uses this framework to provide a detailed analysis of the issues that we have just discussed in the previous paragraph. Section 3 concludes and offers suggestions for future research on slash and burn agriculture.

2. A Model of Land Use in Slash and Burn Agriculture

2.1. *Preliminaries*

The model of this section is adapted from Blanchard (1985). Consider an economy in which small farmers (or swidden cultivators) each have a parcel of cleared forest land and they can choose to grow either a cash crop or a food/subsistence crop on this land. The cash crop requires labor L and high-quality land A_{hq} for production. In contrast, the food or subsistence crop can be grown with labor L and low-quality land A_{lq}. Let w, s_{hq}, and s_{lq} denote the factor rewards to labor, to high-quality land, and to low-quality land, respectively, and let r denote the interest rate. Deleterious and unpredictable environmental events such as earthquakes and volcanoes can make the small farmer's land unfit for cultivation of either the cash crop or the food crop. We account for this possibility by supposing that with instantaneous probability q, $q \in (0, 1)$, the cleared land of our swidden cultivators will become unfit for cultivation. With this contingency in mind, the discount rate of our swidden cultivators is $r + q$.

Our analysis begins with an examination of the decision problem faced by an arbitrary small farmer (or swidden cultivator) at the completion of a specific slash and burn cycle. This corresponds to time $t = 0$ and the quality of the cleared forest land at this point in time is low. Our small farmer must now decide whether to grow the cash crop or the food crop. The food crop can be grown right away because this crop requires labor — which the farmer provides — and low-quality land. In contrast, the cash crop requires labor and high-quality land. Hence, the cash crop cannot be grown immediately.

Now, consistent with the discussion in section 1, our small farmer can convert low-quality land into high-quality land by keeping this land fallow for an appropriate length of time. Put differently, this farmer can choose to accumulate land quality by keeping his or her land fallow. The reader should note two things. First, in the framework of this chapter, optimally choosing the length of time during which the cleared land is to be kept fallow is equivalent to optimally accumulating land quality. Second, the purpose of investing in land quality now is to obtain higher profit from the sale of the cash crop later. Let us now formally study an arbitrary swidden cultivator's decision problem concerning optimal land quality accumulation.

2.2. *Optimal Land Quality Accumulation*

Each swidden cultivator in our economy has a parcel of cleared land that can be left fallow, be used for food crop cultivation right away or — after fallowing — be used for cash crop cultivation. If a parcel of land is left fallow for a time interval of length T, then the relevant small farmer accumulates an amount of land quality equal to $VT^\theta, \theta \in (0, 1]$, where V is a shift variable and θ measures the return to fallowing land.

The key outstanding question now involves the optimal choice of T. To this end, suppose that our arbitrary small farmer wishes to maximize the profit from fallowing land. This profit consists of the earnings from cash crop cultivation less the foregone earnings from food crop cultivation. Now, the earnings from cash crop cultivation are $\int_T^\infty e^{-(r+q)t} VT^\theta s_{hq} dt$, and the foregone earnings from food crop cultivation are $s_{lq}/(r+q)$.[3] Therefore, to determine the optimal length of the fallow period, our small farmer solves

$$\max_{\{T\}} \left[\int_T^\infty e^{-(r+q)t} VT^\theta s_{hq} dt - \frac{s_{lq}}{r+q} \right]. \quad (1)$$

The reader will note that for there to be an interior solution to this problem, there must exist a value of T such that the value of the integral in Equation (1) exceeds the ratio $s_{lq}/(r+q)$. We assume that such an interior solution exists. Now, to determine the first order necessary condition for an optimum, we integrate, using Liebnitz's rule (see Kamien and Schwartz, 1991, p. 292) to carry out the integration. This gives us

$$\frac{\theta VT^{\theta-1} s_{hq}}{r+q} e^{-(r+q)T} - e^{-(r+q)T} VT^\theta s_{hq} = 0. \quad (2)$$

---

3. In the mathematical expression for the earnings from cash crop cultivation, the reader should not interpret the upper limit of integration, which is infinity, literally. We are aware of the fact that it will not be possible for the swidden cultivator to grow the cash crop on his or her land indefinitely after one interval of fallowing. What we mean to suggest with this abstraction is that if the length of the fallow period T is chosen optimally, then it will be possible for our swidden cultivator to grow the cash crop for a relatively long period of time. In addition, the reader should note that keeping the upper limit of integration infinity gives us "clean" mathematical results. Changing the upper limit of integration from infinity to something like $T + G$, where G is the length of time during which the cash crop can be grown without fallowing, is unhelpful because this finite upper limit substantially complicates the mathematical analysis and ultimately leads to ambiguous results. For instance, if the upper limit of integration is $T + G$ then the optimal value of T, T^*, equals $[\theta \exp\{-(r+q)G\} - \theta]/[(r+q)\exp\{-(r+q)G\} - (r+q)]$. To clearly see what we mean by "ambiguous results," the reader should compare this value of T^* with the value of T^* in Equation (3).

The first-order necessary condition in Equation (2) can be simplified to give us the optimal value of T. We get

$$T^* = \frac{\theta}{r+q}. \qquad (3)$$

Equation (3) tells us two things that conform well with our intuition. First, as the return to fallowing land increases, i.e., as 2 gets closer to one, the optimal length of the fallow period — or the optimal length of time during which land quality is accumulated — increases. Second, as our swidden cultivator's discount rate rises, i.e., as $(r+q)$ rises, it is optimal to reduce the length of time during which this cultivator's land is fallow. Inspecting Equation (3), the reader will note that neither the factor reward to high-quality land s_{hq} nor the shift variable V, affect the optimal value of T. This is because s_{hq} and V multiply the term in the profit function that depends on T (see Equation (1)) and hence cancel out of the algebraic simplification process that leads to the derivation of Equation (3). We now analyze some questions that are related to the optimal land quality accumulation issue.

2.3. Related Questions

What are the discounted lifetime earnings that accrue to our swidden cultivator as a result of his or her decision to accumulate land quality optimally? To answer this question, we substitute the optimal value of T from Equation (3) into the integral in Equation (1). This gives

$$\int_{\theta/(r+q)}^{\infty} e^{-(r+q)t} V \left(\frac{\theta}{r+q}\right)^{\theta} s_{hq} dt = e^{-\theta} \frac{1}{r+q} V \left(\frac{\theta}{r+q}\right)^{\theta} s_{hq}. \qquad (4)$$

Now, if there is to be any food crop cultivation in our economy using low-quality land A_{lq}, then, in equilibrium, the swidden cultivator's lifetime earnings in the right-hand side (RHS) of Equation (4) must equal $s_{lq}/(r+q)$, the foregone earnings from food crop cultivation. Setting these two expressions equal to each other gives us an equation for the factor reward of high-quality land relative to that of low-quality land. That equation is

$$\frac{s_{hq}}{s_{lq}} = e^{\theta}\frac{1}{V}\left(\frac{r+q}{\theta}\right)^{\theta}. \qquad (5)$$

The reader will note that in our model, $V(T^*)^{\theta}s_{hq}$, the factor reward for optimally fallowed high-quality land, must exceed s_{lq}, the factor reward of low-quality land.

As we have discussed a little while ago, not all small farmers end up growing the cash crop. In fact, in our model, some swidden cultivators use low-quality land to grow the food crop. Consequently, let us now study the impact that θ, V, r, and q have on s_{lq}, the factor reward to low-quality land. Using Equation (5) and then taking the derivative of s_{lq} with respect to θ, it is easy to see that

$$\frac{ds_{lq}}{d\theta} = s_{lq}\log_e\left(\frac{\theta}{r+q}\right) > 0. \qquad (6)$$

In similar fashion, it can be shown that $ds_{lq}/dV > 0$, that $ds_{lq}/dr < 0$, and that $ds_{lq}/dq < 0$. In words, these results tell us that the factor reward to low-quality land s_{lq} rises with the return to fallowing parameter θ and the shift variable V and it falls with increases in the interest rate r and the probability q of an environmentally disastrous event.

We are now in a position to state a central result about economies with slash and burn agriculture. Specifically, given r, the relative price of the food crop is likely to be higher in economies where there is high demand for keeping land fallow,[4] because of high θ, high V, or low q. Why? As we have already demonstrated, high θ, high V, or low q means that the factor reward for low-quality land s_{lq} is high. In turn, because r is given, a high s_{lq} can be expected to exert an upward pressure on the relative price of the food crop. This concludes our discussion of land use in slash and burn agriculture.

---

4. Indirectly, this implies high demand for the cash crop.

3. Conclusions

We addressed three issues in this chapter that, to the best of our knowledge, have *not* been addressed previously in the theoretical literature on slash and burn agriculture. First, we constructed a theoretical model of land use by shifting cultivators when these cultivators can choose whether to grow a cash crop or a food/subsistence crop. Second, we studied the land quality accumulation decision faced by swidden cultivators and, in the process, we showed how to compute the optimal length of time during which the cleared land is to be left fallow. Finally, we investigated the implications that the optimal land quality accumulation decision has for the factor reward to low-quality land and then we showed that given r, the relative price of the food crop is likely to be higher in economies where there is high demand for keeping land fallow.

The analysis contained in this chapter can be extended in a number of different directions. In what follows, we suggest two avenues for research on the subject of slash and burn agriculture. First, in the model of this chapter, keeping land fallow is a prerequisite for cultivating only the cash crop. As such, it would be useful to analyze a model in which the cultivation of the cash crop and the food crop requires that land be fallowed for a certain length of time. Second, it is well known that different crops affect land quality — and hence the length of the fallow period — in dissimilar ways. Consequently, it would be instructive to analyze a scenario in which small farmers are able to choose between cash crops that have differential impacts on land quality. Studies that incorporate these aspects of the problem into the analysis will provide additional insights into the connections between the land quality accumulation decision and the successful practice of slash and burn agriculture.

References

Batabyal, A.A. and Beladi, H. (2004). Swidden Agriculture in Developing Countries. *Review of Development Economics* 8:255–265.

Blanchard, O.J. (1985). Debt, Deficits, and Finite Horizons. *Journal of Political Economy* 93:223–247.

Brown, K. and Pearce, D.W. (eds.) (1994). *The Causes of Tropical Deforestation*. Vancouver, BC: University of British Columbia Press.

Coomes, O.T., Grimard, F. and Burt, G.J. (2000). Tropical Forests and Shifting Cultivation: Secondary Forest Fallow Dynamics Among Traditional Farmers of the Peruvian Amazon. *Ecological Economics* 32:109–124.

Dove, M.R. (1983). Theories of Swidden Agriculture and the Political Economy of Ignorance. *Agroforestry Systems* 1:85–99.

Dufour, D.L. (1990). Use of Tropical Rainforests by Native Amazonians. *BioScience* 40:652–659.

Ekeleme, F., Akobundu, I.O., Isichei, A.O. and Chikoye, D. (2000). Influence of Fallow Type and Land-Use Intensity on Weed Seed Rain in a Forest/Savanna Transition Zone. *Weed Science* 48:604–612.

Gleave, M.B. (1996). The Length of the Fallow Period in Tropical Fallow Farming Systems: A Discussion with Evidence from Sierra Leone. *Geographical Journal* 162:14–24.

Hofstad, O. (1997). Degradation Processes in Tanzanian Woodlands. *Forum for Development Studies* 0:95–115.

Kamien, M.I. and Schwartz, N.L. (1991). *Dynamic Optimization*, 2nd edn. Amsterdam, The Netherlands: North-Holland.

Li, F., Zhao, S. and Geballe, G.T. (2000). Water Use Patterns and Agronomic Performance for Some Cropping Systems With and Without Fallow Crops in a Semi-Arid Environment of Northwest China. *Agriculture, Ecosystems, and Environment* 79:129–142.

Pearce, D.W. and Warford, J.J. (1993). *World Without End: Economics, Environment, and Sustainable Development*. Oxford, UK: Oxford University Press.

Peters, W.J. and Neuenschwander, L.F. (1988). *Slash and Burn: Farming in the Third World Forest*. Moscow, ID: University of Idaho Press.

Silva-Forsberg, M.C. and Fearnside, P.M. (1997). Brazilian Amazonian Caboclo Agriculture: Effect of Fallow Period on Maize Yield. *Forest Ecology and Management* 97:283–291.

Southgate, D. (1990). The Causes of Land Degradation along Spontaneously Expanding Agricultural Frontiers in the Third World. *Land Economics* 66:93–101.

Swinkels, R.A., Franzel, S., Shepherd, K.D., Ohlsson, E. and Ndufa, J.K. (1997). The Economics of Short Rotation Improved Fallows: Evidence from Areas of High Population Density in Western Kenya. *Agricultural Systems* 55:99–121.

Udaeyo, N.U., Umoh, G.S. and Ekpe, E.O. (2001). Farming Systems in Southeastern Nigeria: Implications for Sustainable Agricultural Production. *Journal of Sustainable Agriculture* 17:75–89.

Chapter 4

A DYNAMIC AND STOCHASTIC ANALYSIS OF FERTILIZER USE IN SWIDDEN AGRICULTURE

With Gregory J. DeAngelo

The number of times a crop can be harvested on a cleared parcel of forest land (CPFL) before this land must be fallowed is dependent on the decision to use or not to use fertilizers to enhance soil fertility. As such, we first construct a theoretical model of fertilizer use by a swidden cultivator when this cultivator can choose whether or not to enhance soil fertility by using fertilizers. Second, we analyze two different policies (fertilizer use and no fertilizer use) for overseeing the problem of soil fertility deterioration on the CPFL. Finally, we identify a particular likelihood function and we show that whether the problem of soil fertility impairment is best addressed with a fertilizer use policy or with a no fertilizer use policy depends essentially on this likelihood function.

1. Introduction

Small-scale farmers in most tropical developing countries practice swidden or slash-and-burn agriculture. There are five key stages in the swidden cycle.[1] First, large forest trees are cut down by the farmer, the debris is cleared, and the cut growth is burned. The burning of the forest vegetation clears the ground for planting and releases vital nutrients. As the burned vegetation decomposes, the

---

[1]. For more on this and related issues, see Dove (1983), Peters and Neunschwander (1988), Pearce and Warford (1993), Brown and Pearce (1994), Swinkels *et al.* (1997), and Coomes *et al.* (2000).

organic levels in the soil rise and this enhances the soil's fertility. Second, before rains cause soil erosion and before the ash bed can be blown or leached away, planting begins. This generally involves the dropping of seeds into shallow holes made by dibble sticks. Third, with the onset of the rainy season, systematic precipitation leads to speedy plant growth. This speedy growth is sometimes accompanied by the concurrent growth of weeds. These weeds are regularly removed by the swidden cultivator to preclude them from taking nutrients away from the crop under cultivation. Fourth, during the harvesting season, the cultivator protects the crop from pests and (s)he often uses simple implements to harvest the grain. In the process of harvesting the grain, the cultivator retains some of the best seeds for the next planting. Finally, and this is the crucial stage, the cleared parcel of forest land (CPFL) is left fallow after one or two harvests. Within a couple of years, a closed canopy of secondary forest develops. If the CPFL is left fallow for an adequately long period of time then nutrients will return to the soil and this will permit the above described swidden cycle to be repeated.

Despite the significance of swidden agriculture in tropical developing countries, there is some controversy about the merits of this kind of agriculture. On one hand, scholars such as Dove (1983), Southgate (1990), and Pearce and Warford (1993) have criticized this kind of agriculture. Specifically, these scholars have claimed that swidden agriculture is environmentally harmful because the land clearing activities of swidden cultivators is directly associated with extensive and ruinous tropical forest deforestation. On the other hand, a second group of scholars including Peters and Neuenschwander (1988) and Dufour (1990) have argued that in some situations, swidden agriculture based on long fallow periods can be an ecologically and an economically sustainable enterprise in tropical forests.

Batabyal and Lee (2003) and Batabyal and Beladi (2004) have recently argued convincingly that the viability of swidden agriculture in the long run depends on a CPFL being fallowed at appropriate points in time and for appropriate lengths of time.[2] Although this is certainly true, as noted by Dickinson (1972), Farnsworth and Golley (1973), and Eckholm (1976), what is also true is that swidden cultivators can — and often have attempted to — increase the number of harvests on a particular CPFL before this CPFL must be fallowed by applying natural and/or chemical fertilizers. However, beyond recognizing this essential point, scholars have not *theoretically* analyzed the fertilizer use decision problem faced by swidden cultivators. In addition, keeping in mind the dynamic and the stochastic setting in which swidden cultivators typically operate, scholars have *not* studied the conditions under which it is optimal to use fertilizers.

Given this state of affairs, our chapter has three goals. Section 2 describes a dynamic and stochastic model of fertilizer use. Section 3 first analyzes two different policies (fertilizer use and no fertilizer use) for overseeing the problem of soil fertility deterioration. Next, this section identifies a specific likelihood function and it shows that in addition to cost considerations, whether the problem of soil fertility deterioration is best addressed with a fertilizer use policy or a no fertilizer use policy depends primarily on this likelihood function. Finally, Section 4 concludes and discusses ways in which the research of this chapter might be extended.

2. The Theoretical Framework

This chapter's model is adapted from previous research by Antelman and Savage (1965), Batabyal and Yoo (1994), and Ross (1996, Chapter 8). Consider a swidden cultivator and a dynamic and stochastic parcel of forest land that has just been cleared for the planting of a particular crop. As explained in Section 1, this cultivator follows the swidden cycle and thereby obtains grain output from

---

2. For a more expansive discussion of this and related points, see Hofstad (1997), Silva-Forsberg and Fearnside (1997), Coomes *et al.* (2000), Li *et al.* (2000), and Udaeyo *et al.* (2001).

successive harvests of the crop under study. Now, *ceteris paribus*, our cultivator would like to repeat as many swidden cycles as possible on the CPFL but in doing this, (s)he must contend with the deterioration in soil fertility on this CPFL. Put differently, our cultivator realizes that after a certain number of crop harvests, the CPFL will not yield any tangible grain output and hence this CPFL will need to be fallowed.

It is generally possible to extend the useful life of a CPFL, i.e., to increase the number of harvests before the land is fallowed, by using fertilizers.[3] This is a rationale for a policy of fertilizer use by our cultivator. However, as noted by Dickinson (1972), Eckholm (1976), and others, fertilizers are costly — chemical fertilizers more so than natural ones — and because swidden cultivators are typically poor, small-scale farmers, they often will not possess the financial resources to use fertilizers. In other words, although fertilizer use can be economically profitable, the cost of fertilizer use frequently serves as a deterrent against its use.

We can now state the two policies that are available to our swidden cultivator. The first policy is a passive one in which no fertilizer is used. Clearly, when our cultivator uses this policy, (s)he is relying solely on natural or environmental factors to delay the deterioration in soil fertility. In contrast, the second policy is an active one in which the cultivator uses fertilizers and thereby actively attempts to retard the deterioration in soil fertility.

The essential stock variable that is affected by the repeated planting of the crop in question is the stock of soil fertility. Notionally, this stock is very much like the stock of an exhaustible natural resource such as oil. Just as repeated extraction of this exhaustible resource draws down its stock, similarly, repeated planting of a crop lowers the stock of soil fertility. The reader should note that because of a variety of reasons, not the least of which is the swidden cultivator's own crop planting actions, the lowering of the stock of soil fertility

---

3. In the rest of this chapter, when we refer to fertilizer use, we are referring to both natural and chemical fertilizer use.

is typically *probabilistic* and not deterministic. To account for this aspect of the problem, we shall think of the soil fertility stock as a stochastic process that can exist in one of many possible states. To this end, let state 0 be the best possible state of the soil fertility stock. The reader should think of this state as corresponding to the state in which our CPFL exists immediately after the completion of the first stage in the five stage swidden cycle that we described in Section 1. To model the stochastic soil fertility degradation process, we shall say that the stock of soil fertility changes state in accordance with a Wiener process with drift $\delta > 0$.[4]

With repeated planting of the crop in question, soil fertility on the CPFL deteriorates, our Wiener process changes state, and eventually this process gets to a "breakdown" state in which the land must be fallowed. Denote this breakdown or *fallow* state by f. The idea here is that once this fallow state is reached, our swidden cultivator's reliance on the passive policy in which natural or environmental factors alone retard the deterioration in soil fertility, has run its course. Consequently, the CPFL must now be left fallow for a certain period of time.[5] When this is done, a secondary forest cover gradually emerges on the CPFL. Mathematically, this means that our Wiener process eventually returns to state 0. In other words, the state of the stock of soil fertility corresponds, once again, to the state in which the CPFL exists right after the completion of the first stage in the five stage swidden cycle. When the fallow state is reached, our cultivator must abandon the CPFL under study and this cultivator must now look for and clear an alternate parcel of forest land. These land clearing activities are costly. As such, let us denote the cost of the passive or no fertilizer use policy by $c(f)$.

We now focus on the active policy in which our swidden cultivator uses fertilizers. The reader should note that we are using the

---

4. For more on the Wiener process, see Ross (1996, Chapter 8; 2003, Chapter 10).

5. An interesting question that emerges in this context is the determination of the optimal length of time during which the CPFL ought to be fallow. This question is addressed in Batabyal and Beladi (2004).

word "fertilizer" in a general way. In particular, this active policy may involve the use of a natural or a chemical fertilizer. Further, depending on the crop that is being grown, it may even make sense to use more than one fertilizer. The key thing to note is that in contrast with the previously described passive or no fertilizer use policy, the active policy always involves fertilizer use. Finally, note that for the active policy to make sense, our swidden cultivator must use the fertilizer(s) before the Wiener process hits the fallow state f.

Now, because our swidden cultivator operates in a probabilistic environment, the pursuit of the fertilizer use policy increases the likelihood that soil fertility on the CPFL will improve but it does *not* guarantee that soil fertility will improve. Specifically, if the state of the Wiener process is b and the active or fertilizer use policy is utilized, then this policy will be successful in improving soil fertility with probability $p(b)$, and it will be unsuccessful with probability $1 - p(b)$. Why might a fertilizer use policy be unsuccessful in raising soil fertility? Two possible reasons come to mind. First, our swidden cultivator may use too little or too much fertilizer and, as a result, soil fertility might not be impacted in any significant manner. Second, this cultivator may use the wrong fertilizer and hence, once again, there will be no improvement in soil fertility. If the fertilizer use policy is successful in improving soil fertility then we suppose that our Wiener process returns to state 0. In other words, the state of the stock of soil fertility corresponds to the state in which the CPFL exists right after the completion of the first stage in the five stage swidden cycle. In contrast, if this fertilizer use policy is unsuccessful in improving soil fertility then the Wiener process is assumed to go to state f.[6] The cost of attempting to improve soil fertility actively in state b is $c(b)$. Our charge now is to ascertain which policy, active or passive, minimizes the long-run average cost per time.

---

6. We understand that the failure of the fertilizer use policy does not necessarily mean that soil fertility has declined to such an extent that our Wiener process must go to state f. We make this assumption primarily for reasons of mathematical tractability. Having said this, we recognize that it is possible that the Wiener process will go to some intermediate state e, where e is worse than b but better than state f.

3. Active Versus Passive Policies

3.1. Long-Run Average Cost per Time

To calculate this cost function, we shall use renewal theory (see Ross 1996, Chapter 3; 2003, Chapter 7). Further, as far as the fertilizer use policy is concerned, we shall focus attention on those policies that attempt to raise soil fertility when our Wiener process is in state b, where $0 < b < f$. Given this stipulation, note that every time our Wiener process returns to state 0, we have a renewal. Consequently, we can use a prominent result in renewal theory, namely, the renewal-reward theorem,[7] to calculate the long-run average cost that we seek. Now, if we envision a cycle being completed every time a renewal occurs, then the renewal-reward theorem tells us that the long-run average reward or cost (a negative reward) is given by the expected positive or negative return earned in a cycle divided by the length of this cycle.

We now adapt the renewal-reward theorem to fit the problem that we are analyzing. This gives us

$$\text{Long-Run Average Cost} = \frac{E[\text{Cost per Cycle}]}{E[\text{Length of Cycle}]}, \quad (1)$$

where $E[\cdot]$ is the expectation operator. Now, calculating the numerator on the right-hand side (RHS) of Equation (1) is fairly routine. Some thought tells us that the average cost per cycle is given by $c(b) + \{1 - p(b)\}c(f)$. Hence, in symbols, we have

$$E[\text{Cost per Cycle}] = c(b) + \{1 - p(b)\}c(f). \quad (2)$$

The calculation of the average length of a renewal cycle is more involved. We proceed as in Batabyal and Yoo (1994). Let us represent the average time it takes for our Wiener process to reach state b with the function $k(b)$. Now, it is well known that a Wiener process has independent and stationary increments.[8] Therefore, for any two

---

7. For more on the renewal-reward theorem, see Ross (1996, p. 133) or Ross (2003, p. 417).
8. For more on these notions, see Ross (1996, Chapter 8; 2003, Chapter 10).

states of the process b_1 and b_2, we can write

$$k(b_1 + b_2) = k(b_1) + k(b_2). \tag{3}$$

Now, the previously mentioned properties of Wiener processes and Equation (3) together tell us that the function $k(b)$ has the form $k(b) = z \cdot b$, where z is a constant. It can be demonstrated that the constant $z = 1/\delta$ and, therefore, $k(b) = b/\delta$.[9] This last finding permits us to conclude that

$$E[\text{Length of Cycle}] = \frac{b}{\delta}. \tag{4}$$

Now, using Equations (2) and (4) together, we get

$$[\text{Long-Run Average Cost}]_A = \frac{E[\text{Cost per Cycle}]}{E[\text{Length of Cycle}]}$$
$$= \frac{\delta[c(b) + \{1 - p(b)\}c(f)]}{b}. \tag{5}$$

Equation (5) tells us that the long-run average cost of raising soil fertility with the active or fertilizer use policy is given by the ratio of the weighted sum of the two cost expressions $c(b)$ and $c(f)$ to the state $b, 0 < b < f$, in which this policy is utilized.

Our charge now is to ascertain the long-run average cost of the passive or no fertilizer use policy. For this policy, it is easy to see that $E[\text{Cost of Cycle}] = c(f)$. Further, following the logic of the derivation that led to Equation (4), we deduce that $E[\text{Length of Cycle}] = f/\delta$. Hence, putting these two pieces of information together, we reason that

$$[\text{Long-Run Average Cost}]_P = \frac{E[\text{Cost per Cycle}]}{E[\text{Length of Cycle}]} = \frac{\delta c(f)}{f}. \tag{6}$$

According to Equation (6), the long-run average cost of the passive policy in which our swidden cultivator relies exclusively on natural or environmental factors to improve soil fertility is given by the ratio

---

9. For additional details on this point, see Batabyal and Yoo (1994) and Ross (1996, Chapter 8).

of the product of the drift parameter of our Wiener process δ and the cost of locating and clearing an alternate parcel of forest land $c(f)$ to the fallow state f.

Examining Equation (5) it should be clear to the reader that for a specific likelihood function $p(b)$, we can always use calculus to minimize this long-run average cost function. Even so, we now discuss an important point and that point is this: The choice between the active policy and the passive policy (see Equations (5) and (6)) is, in large part, contingent on the likelihood function $p(b)$.

3.2. Significance of the Likelihood Function

To see the above point, let us first equate the RHS of Equations (5) and (6). This gives us a threshold value for the likelihood function and that value is

$$p(b) = \frac{c(b)}{c(f)} - \frac{b}{f} + 1. \qquad (7)$$

When the likelihood function equals the RHS of Equation (7), our swidden cultivator will be indifferent between pursuing the fertilizer use policy and the no fertilizer use policy because both policies give rise to the same long-run average cost.

In contrast, consider the case where $p(b) = 1 - b/f$. In this case $1 - p(b) = b/f$, and substituting this value into the RHS of Equation (5) and then comparing the result with the RHS of Equation (6), we see that

$$[\text{Long-Run Average Cost}]_A = \frac{\delta c(b)}{b} + \frac{\delta c(f)}{f} > \frac{\delta c(f)}{f}$$

$$= [\text{Long-Run Average Cost}]_P. \qquad (8)$$

Equation (8) explicitly tells us that when the likelihood function is $p(b) = 1 - b/f$, the optimal course of action for our swidden cultivator is to not use fertilizers on the CPFL. This cultivator does better, i.e., bears a lower long-run average cost by relying solely on natural or environmental factors to improve soil fertility on the CPFL.

As a second illustration, consider the likelihood function $p(b) = \{c(b)/c(f)\} - (b/f) + 2$. In this case $1 - p(b) = (b/f) - \{c(b)/c(f)\} - 1$ and, once again, substituting this value into the RHS of Equation (5) and then comparing the result with the RHS of Equation (6), we see that

$$[\text{Long-Run Average Cost}]_A = \frac{\delta c(f)}{f} - \frac{\delta c(f)}{b} < \frac{\delta c(f)}{f}$$

$$= [\text{Long-Run Average Cost}]_P. \quad (9)$$

In this second illustration, Equation (9) clearly tells us that when the likelihood function is $p(b) = \{c(b)/c(f)\} - (b/f) + 2$. it is optimal for our swidden cultivator to pursue the active policy in which (s)he relies on fertilizers — and not on natural or environmental factors — to improve soil fertility on the CPFL. Put differently, this cultivator bears a lower long-run average cost by using fertilizers to raise soil fertility on the CPFL.

4. Conclusions

In this chapter, we provided a theoretical analysis of the "to use or not to use fertilizer" question facing a swidden cultivator. In particular, we shed light on two issues. First, we analyzed two different policies (fertilizer use and no fertilizer use) for overseeing the problem of soil fertility deterioration. Second, we identified a particular likelihood function and we showed that whether the problem of soil fertility impairment is best addressed with a fertilizer use policy or with a no fertilizer use policy depends essentially on this likelihood function.

The analysis contained in this chapter can be extended in a number of ways. In what follows, we propose two potential extensions of this chapter's research. First, the reader will note that we modeled the stochastic decline in soil fertility on the CPFL with a Wiener process. Therefore, it would be interesting to examine the extent to which the results of this chapter hold when alternate stochastic processes are used to model the random movement toward the fallow state. Second, it would be useful to compare the approach of this chapter with an alternate approach in which in addition to the fallowing costs and the costs of fertilizer use, it is costly for a swidden cultivator to determine the current state of the stochastic process denoting soil fertility. Studies of fertilizer use in swidden agriculture that incorporate these aspects of the problem into the analysis will provide additional insights into how one might manage the problem of soil fertility deterioration.

References

Antelman, G. and Savage, I.R. (1965). Surveillance Problems: Wiener Processes. *Naval Research Logistics Quarterly* 12:35–55.

Batabyal, A.A. and Yoo, S.J. (1994). Renewal Theory and Natural Resource Regulatory Policy Under Uncertainty. *Economics Letters* 46:237–241.

Batabyal, A.A. and Lee, D.M. (2003). Aspects of Land Use in Slash and Burn Agriculture. *Applied Economics Letters* 10:821–824.

Batabyal, A.A. and Beladi, H. (2004). Swidden Agriculture in Developing Countries. *Review of Development Economics* 8:255–265.

Brown, K. and Pearce, D.W. (eds.) (1994). *The Causes of Tropical Deforestation*. Vancouver, BC: University of British Columbia Press.

Coomes, O.T., Grimard, F. and Burt, G.J. (2000). Tropical Forests and Shifting Cultivation: Secondary Forest Fallow Dynamics Among Traditional Farmers of the Peruvian Amazon. *Ecological Economics* 32:109–124.

Dickinson, J.C. (1972). Alternatives to Monoculture in the Humid Tropics of Latin America. *Professional Geographer* 24:217–222.

Dove, M.R. (1983). Theories of Swidden Agriculture and the Political Economy of Ignorance. *Agroforestry Systems* 1:85–99.

Dufour, D.L. (1990). Use of Tropical Rainforests by Native Amazonians. *BioScience* 40:652–659.

Eckholm, E.P. (1976). *Losing Ground*. New York: W.W. Norton.

Farnsworth, E.G. and Golley, F.B. (eds.) (1973). *Fragile Ecosystems*. New York: Springer-Verlag.

Hofstad, O. (1997). Degradation Processes in Tanzanian Woodlands. *Forum for Development Studies* 0:95–115.

Li, F., Zhao, S. and Geballe, G.T. (2000). Water Use Patterns and Agronomic Performance for Some Cropping Systems With and Without Fallow Crops

in a Semi-Arid Environment of Northwest China. *Agriculture, Ecosystems, and Environment* 79:129–142.
Pearce, D.W. and Warford, J.J. (1993). *World Without End: Economics, Environment, and Sustainable Development.* Oxford, UK: Oxford University Press.
Peters, W.J. and Neuenschwander, L.F. (1988). *Slash and Burn: Farming in the Third World Forest.* Moscow, ID: University of Idaho Press.
Ross, S.M. (1996). *Stochastic Processes*, 2nd edn. New York, NY: Wiley.
Ross, S.M. (2003). *Introduction to Probability Models*, 8th edn. San Diego, CA: Academic Press.
Silva-Forsberg, M.C. and Fearnside, P.M. (1997). Brazilian Amazonian Caboclo Agriculture: Effect of Fallow Period on Maize Yield. *Forest Ecology and Management* 97:283–291.
Southgate, D. (1990). The Causes of Land Degradation along Spontaneously Expanding Agricultural Frontiers in the Third World. *Land Economics* 66:93–101.
Swinkels, R.A., Franzel, S., Shepherd, K.D., Ohlsson, E. and Ndufa, J.K. (1997). The Economics of Short Rotation Improved Fallows: Evidence from Areas of High Population Density in Western Kenya. *Agricultural Systems* 55:99–121.
Udaeyo, N.U., Umoh, G.S. and Ekpe, E.O. (2001). Farming Systems in Southeastern Nigeria: Implications for Sustainable Agricultural Production. *Journal of Sustainable Agriculture* 17:75–89.

Chapter 5

ON FLOOD OCCURRENCE AND THE PROVISION OF SAFE DRINKING WATER IN DEVELOPING COUNTRIES

Developing countries in South Asia and elsewhere are frequently ravaged by floods. An important part of most flood management programs is the provision of safe drinking water (SDW) to prevent the spread of diseases. How should a government agency that is interested in distributing SDW to flood victims, go about its task? Further, how might this agency maximize the net social benefit from the provision of SDW? Finally, given that SDW is a particularly scarce commodity in a flood situation, how likely is it that this agency will be unable to meet the stochastic demand for SDW? In this chapter, we use queuing theory to shed light on these three questions regarding the disbursement of SDW to flood victims.

1. Introduction

In contemporary times, approximately one billion people in the developing world do not have access to safe drinking water (SDW) (Balint, 1999). Given this state of affairs, an important question with significant public policy implications concerns the provision of adequate and SDW for the burgeoning populations of the world's developing countries. This question is important because SDW is essential for preventing a whole host of water borne diseases. As well, access to SDW ensures a minimal level of health and bodily health is a central human capability (Nussbaum, 2000, p. 78).[1]

A recognition of these facts has led national and international development agencies to embark on a whole host of schemes to provide SDW in developing countries.[2] Now, many of the world's developing nations, particularly those in South Asia, are frequently ravaged by floods. Further, as noted in Emch (2000), the unavailability of SDW is the proximate cause of a variety of water borne diseases in these flood prone areas. As such, when a flood occurs, the question of providing SDW to flood victims assumes particular salience.

How should a government agency that is interested in distributing SDW to flood victims, go about its task? Further, how might this agency maximize the net social benefit from the provision of SDW? Finally, given that SDW is a particularly scarce commodity in a flood situation, how likely is it that this agency will be unable to meet the stochastic demand for SDW? Although there are a number of empirical and case study based analyses of drinking water problems in developing countries (see Han *et al.* (1991), Asthana (1997), Balint (1999), and Reddy (1999)), and even some studies of drinking water problems in flood situations (see Haque and Zaman (1993) and Emch (2000)), to the best of our knowledge, there are *no theoretical* studies of the three questions that we posed earlier in this paragraph.

As such, the purpose of this chapter is to show how queuing theory can be used to effectively model and study these three questions concerning the disbursement of SDW in flood prone developing countries. The rest of this chapter is organized as follows: Section 2 describes the queuing theoretic model in detail. Section 3 first discusses aspects of this model and then analyzes an optimization problem for our SDW disbursing government agency. Section 4 concludes and offers suggestions for future research.

---

1. For an expansive discussion of capabilities, see Sen (1985; 1992), Nussbaum (2000), and the many references cited in these three books.

2. For more on this, see Munasinghe (1992), Balint (1999), and Kleemeier (2000).

2. The Queuing Theoretic Approach to the Provision of SDW

Batabyal (1996) was the first to demonstrate the applicability of queuing theory to water allocation problems. Although Batabyal's (1996) paper analyzed the allocation of groundwater *per se*, in what follows, we shall use the methods developed in that paper to analyze the three questions posed in the previous section.

Consider a state like West Bengal in a developing country such as India that has been affected by a flood.[3] There are many flood victims and the government agency that is of interest to us has been entrusted with the task of providing SDW to these flood victims.[4] To this end, we suppose that this agency sets up a relief center. The agency imports SDW to this center by means of tanker trucks and it "produces" buckets of SDW in accordance with a Poisson process whose rate is $\lambda > 0$. Because SDW is a scarce commodity, there is a limit to how many buckets of water our agency can produce during a given time period, say, a day. To model this, we shall say that once K buckets of water have been produced, the agency has exhausted its available supply of SDW and hence it must wait for more water to appear.

Suppose that flood victims arrive at this relief center in accordance with a Poisson process with rate $\mu > 0$. Each victim is entitled to one bucket of water per day. Upon receipt of this bucket, the victim leaves the relief center. If this victim happens to arrive at the center when buckets of water have already been disbursed, then (s)he will have to leave the center empty handed.

Formally, the events described in the previous two paragraphs can be modeled effectively with the $M/M/1$ queue with finite capacity

---

3. West Bengal is frequently ravaged by floods. As reported in *The Economist* (2000, p. 6), in year 2000 alone, upwards of 17 million people have been adversely affected by floods in this state and in neighboring Bangladesh.

4. Government agencies assigned the task of flood control typically perform many duties, only one of which is the provision of SDW. We have focused on SDW in this chapter because of the fundamental importance of SDW in sustaining human life and because we wish to shed light on hitherto unanswered research questions in flood management.

K.[5] In this three-part classification, the first M refers to the fact that the times between successive productions of buckets of SDW by the relief center have the Markovian property. The second M refers to the times between the arrivals of successive flood victims to the relief center; these times also have the Markovian property. Finally, the 1 in the three-part classification scheme refers to the fact that our analysis involves only one relief center. Let us now investigate this queuing theoretic model in greater detail.

3. Analysis

Our first task is to define the state space for our queue. To this end, let $X(t)$ denote the number of buckets of SDW that are available at time t in the relief center. We can now express the stationary or limiting probabilities for our queue. We get

$$P_k = \lim_{t \to \infty} \text{Prob}\{X(t) = k\}, \quad k = 0, 1, \ldots, K, \qquad (1)$$

where the subscript k denotes the buckets of SDW that are currently available for disbursement. By Proposition 8.2 in Ross (2000, p. 432), these stationary probabilities, i.e., the P_k's also denote the proportion of time that the buckets of SDW are in a particular state k, $k = 0, \ldots, K$.

Given the public service nature of the government agency's mission, one ratio that this agency is interested in is the proportion of all flood victims who find the relief center filled with buckets of SDW. To compute this ratio, we shall use the discussion in the previous paragraph. Using this discussion, the proportion of interest is equal to the limiting probability (P_K) that the relief center has K buckets of SDW. Now, from Equation (7) in Batabyal (1996, p. 222), we conclude that P_K satisfies

$$P_K = \frac{(\lambda/\mu)^K \{1 - (\lambda/\mu)\}}{1 - (\lambda/\mu)^{K+1}}. \qquad (2)$$

---

5. For more details on the $M/M/1$ queue with finite capacity, see Batabyal (1996) and Ross (2000, pp. 432–447).

Having determined this probability, we are now in a position to formulate an optimization problem for our government agency. In our queuing theoretic context, there are several reasonable objectives that we could set up for the government agency. One such objective is the culmination of the following line of reasoning: First, evaluate the benefit from SDW provision. In the time period under consideration, i.e., a day, our agency obtains social benefit B^* from SDW provision. We suppose that $B^* = B\mu$. In other words, the social benefit is a function of the rate (μ) at which SDW is provided to flood victims and B is the constant marginal social benefit from the provision of SDW.

Next consider the cost of SDW provision. To this end, suppose that our agency incurs both fixed and variable costs. The fixed social costs are $\$F$ per day. The variable social costs are a little more involved. To see this, first recall that there is a capacity K on the daily production of SDW for flood victims. As such, the stationary probability P_K tells us the proportion of time that the relief center is full with buckets of SDW. From this we conclude that the rate at which SDW is made available to flood victims is $\lambda(1 - P_K)$ per day. Finally, let $\$V$ denote the variable cost of providing buckets of SDW per day. Then, the daily variable social cost is $\lambda V(1 - P_K)$.

We are now in a position to state our government agency's objective function and its optimization problem. This objective function is the net social benefit function $B\mu - \{\lambda V(1 - P_K) + F\}$, where P_K is given by Equation (2), and this agency solves

$$\max_{\{\mu\}} B\mu - \frac{\lambda V\{1 - (\lambda/\mu)^K\}}{1 - (\lambda/\mu)^{K+1}} - F, \qquad (3)$$

where the control variable μ is the rate at which SDW is provided to flood victims. The first-order necessary condition for an interior maximum is[6]

$$B = \frac{\{1 - (\lambda/\mu)^{K+1}\}\{VK(\lambda/\mu)^{K+1}\} - \{\lambda V - \lambda^{K+1} V\mu^{-K}\}\{(K+1)\lambda^{K+2}\mu^{-(K+3)}\}}{1 - (\lambda/\mu)^{K+1}}.$$

$$(4)$$

---

6. We assume that the second order condition is satisfied.

Equation (4) tells us that the optimal rate of SDW provision should be chosen so that the marginal social benefit from water provision (the LHS) is equal to the marginal social cost (the RHS). In general, it will not be possible to solve Equation (4) for the optimal μ explicitly. As such, it will be necessary to resort to numerical methods to determine the optimal rate of SDW provision.

In Section 1, we noted that one question that has not been addressed previously in the literature concerns the *likelihood* that a government agency of the sort considered in this chapter will be unable to meet the stochastic demand for SDW in a flood situation. We conclude this section by showing how this question can be answered. First, we will need to ascertain the proportion of flood victims who come to the agency's relief center and are unable to obtain SDW from the agency. Now, the agency will be unable to provide SDW to a flood victim only if this victim arrives at the relief center after the agency has run out of SDW for that day. This tells us that the proportion that we are after (also see Equation (1)) is actually the limiting probability $P_0 = \lim_{t \to \infty} \text{Prob}\{X(t) = 0\}$. From Equation (6) in Batabyal (1996, p. 222), we reason that

$$P_0 = \frac{1 - (\lambda/\mu)}{1 - (\lambda/\mu)^{K+1}}. \tag{5}$$

We see that this probability depends on the parameters of the queue (λ, μ) and on the capacity of the relief center (K) for providing SDW. Further, in the general case, an increase in μ, the rate at which SDW is provided to flood victims, has an ambiguous impact on the proportion of flood victims who come to the agency's relief center and are unable to obtain SDW.

4. Conclusions

In this chapter, we analyzed a simple model of the provision of SDW in a flood situation in developing countries. Specifically, we showed how queuing theory can be used to model and study the optimization problem faced by a government agency that is responsible for disbursing SDW to flood victims. Our analysis leads to two conclusions. First, the operator of the relief center, i.e., the government agency, should disburse SDW to flood victims at a rate such that the marginal social benefit from the provision of water equals the marginal social cost of providing water. Second, Equation (5) showed us that it is possible to quantify the likelihood that the relief center will be unable to meet the stochastic demand for SDW by flood victims.

The analysis of this chapter can be extended in a number of directions. In what follows, we suggest two possible extensions. First, one could analyze the SDW disbursement problem in a queuing theoretic framework in which either the SDW production times or the SDW provision times are generally (and not exponentially) distributed. Second, it would be interesting to compare the solution to this chapter's optimization problem (Equation (3)) with the solution to an alternate problem that involves the minimization of a function of the stationary probability described in Equation (5). Studies of the SDW disbursement problem that incorporate these features of the problem into the analysis will provide additional insights into the role that effective water provision strategies play in flood control operations.

References

Anonymous (2000). Politics this Week. *The Economist* September 30:6.

Asthana, A.N. (1997). Where the Water is Free but the Buckets are Empty: Demand Analysis of Drinking Water in Rural India. *Open Economies Review* 8:137–149.

Balint, P.J. (1999). Drinking Water and Sanitation in the Developing World: The Miskito Coast of Nicaragua and Honduras as a Case Study. *Journal of Public and International Affairs* 10:99–117.

Batabyal, A.A. (1996). The Queuing Theoretic Approach to Groundwater Management. *Ecological Modelling* 85:219–227.

Emch, M. (2000). Relationships Between Flood Control, Kala-azar, and Diarrheal Disease in Bangladesh. *Environment and Planning A* 32:1051–1063.

Han, G., Jiang, F. and Yan, J. (1991). 2000AD: Water Environment Problems of China. *International Journal of Social Economics* 18:174–179.

Haque, C.E. and Zaman, M.Q. (1993). Human Responses to Riverine Hazards in Bangladesh: A Proposal for Sustainable Floodplain Development. *World Development* 21:93–107.

Kleemeier, E. (2000). The Impact of Participation on Sustainability: An Analysis of the Malawi Rural Piped Scheme Program. *World Development* 28:929–944.

Munasinghe, M. (1992). *Water Supply and Environmental Management: Developing World Applications.* Boulder, CO: Westview Press.

Nussbaum, M.C. (2000). *Women and Human Development.* Cambridge, UK: Cambridge University Press.

Reddy, V.R. (1999). Pricing of Rural Drinking Water: A Study of Willingness and Ability to Pay in Western India. *Journal of Social and Economic Development* 2:101–122.

Ross, S.M. (2000). *Introduction to Probability Models*, 7th edn. San Diego, CA: Academic Press.

Sen, A.K. (1985). *Commodities and Capabilities.* Amsterdam, The Netherlands: North-Holland.

Sen, A.K. (1992). *Inequality Reexamined.* Cambridge, MA: Harvard University Press.

Chapter 6

RENEWABLE RESOURCE MANAGEMENT IN DEVELOPING COUNTRIES: HOW LONG UNTIL CRISIS?

With Hamid Beladi

A key goal of renewable resource managers in developing countries is to take actions to ensure that the resource being managed stays away from irreversible or crisis states in which it provides neither consumptive nor non-consumptive services to humans. However, despite a manager's best efforts, the resource may still hit a crisis state. Therefore, given a particular management regime, it is useful to know how long it takes until the resource hits a crisis state. In this chapter, we provide a theoretical analysis of this hitherto unstudied question. We first probabilistically delineate two management regimes. Next, we compute the expected time until crisis for both these regimes. Finally, we provide a numerical example to illustrate the working of our model and then we discuss the implications of our findings for renewable resource management in developing countries.

1. Introduction

Extensive contemporary discussions about sustainable development have brought the subject of renewable resource management into sharp focus (Dasgupta and Maler, 1997). Although there is some disagreement among researchers about the meaning of the term "sustainable development," there is general agreement on the point that if the process of development is to be sustainable, then essential renewable resources such as fisheries, forests, groundwater, and rangelands will need to be optimally managed.

People in developing countries are largely agrarian and pastoral in nature. In other words, they are significantly *dependent* on agriculture and on renewable resources, particularly those renewable resources that are found in their local environment.[1] For instance, villagers living in the watershed of the Alaknanda River in the central Himalayas in India spent 20% of their total working time collecting fodder and 25% of their total working time collecting fuel, caring for animals, and grazing their animals (Center for Science and Environment, 1990). This dependence on renewable resources is not limited to South Asia. Consider the case of Africa. Falconer and Arnold (1989) and Falconer (1990) have shown that forests and the associated forest products are vital to the lives of rural people in Central and West Africa. Given the nature and the extent of this dependence on renewable resources, it should be clear to the reader that the optimal management of these resources is an extremely important policy objective. Indeed, in this context, it is useful to point out that environmental problems are "almost always associated with resources that are regenerative but are in danger of exhaustion from excessive use" (Dasgupta, 1996, p. 389).

Renewable resources can exist in a number of different states. To fix ideas, consider the following binary state descriptions of two resources: A prudently exploited fishery can exist in a "desirable" state in which the fish stock is high; in contrast, if overexploited, then this fishery will most likely revert to an "undesirable" state in which the fish stock is low.[2] Similarly, rangelands that have not been overgrazed can exist in a "desirable" state in which there is substantial forage cover. On the other hand, it is now well known that overgrazed rangelands can revert to an "undesirable" desertified state.[3]

Generalizing from the above two examples, a renewable resource can reasonably be thought to exist in not just two but a finite number

---

1. See Agarwal (2000) for an interesting account — with examples from rural South Asia — of how this dependence on renewable resources can be gender based as well.

2. See Clark (1990) for an excellent general account of issues in the optimal management of fisheries.

3. For more on these issues, see Jodha (1980), Ho (2000), and Batabyal and Godfrey (2002).

of states. Some of these states are desirable and others are undesirable. Also, the reader should note that in both these sets of states, some states are better than others. Restricting attention to the undesirable set of states, what is important for our purpose is that some states are likely to be *irreversible*. In these irreversible or *crisis* states,[4] the resource is so degraded — the fish stock is very low or forage cover on a rangeland is virtually non-existent — that no matter how hard a manager might try, (s)he will be unable to move the resource to any other state. Given that this is the case, one way to look at the task of renewable resource management is to say that a manager's[5] objective is to maximize the amount of time a resource spends in the desirable set of states. Using the same logic, one can also say that a resource manager's task is to minimize the amount of time the resource spends in the undesirable — but not the irreversible — set of states.

Now, all resource managers operate in a *stochastic* environment. In other words, the state of a managed resource at any particular point in time is a function not only of the actions undertaken by this manager but also of unpredictable environmental factors like droughts, fires, and predators. What this means is that even though the manager believes that (s)he is taking actions to ensure that, at the very least, a resource does not hit any one of the crisis states, the resource may still do so. Put differently, despite the manager's best efforts, (s)he can never be certain that the managed resource will not hit an irreversible state.

Given this state of affairs, how should a developing country resource manager proceed? We now study the properties of the following reasonable approach: First, identify the crisis state or states. Although it is generally not possible to do so precisely, on the basis of past experience and available scientific information, managers are

---

4. In the rest of this chapter, we shall use the terms "irreversible state" and "crisis state" interchangeably.

5. The manager need not be a single individual. In many developing countries, communities collectively manage renewable resources. For more on this, see Wade (1988) and Dasgupta (1996).

generally able to specify a safe minimum standard,[6] and all resource states below the safe minimum standard can be thought of as crisis states. Next, it is necessary to put in place a well-designed plan of action. In the context of fisheries, this might mean a combination of season length and gear restrictions and in the context of rangelands, this might mean the use of a specialized grazing system such as short duration grazing.[7]

As discussed in the previous two paragraphs, even with a well-designed plan of action in place, it is still possible that a managed resource will hit a crisis state. Consequently, a key question now is this: How long until crisis? In other words, the manager would like to know how long it takes for the resource to hit a crisis state. Further, what does the answer to this question depend on? Finally, is the answer history dependent or independent? In other words, does the answer to the how long until crisis question depend on the state in which the manager's plan of action is put in place? Clearly, answers to these questions are vital for successful renewable resource management in developing countries. Yet, to the best of our knowledge, there are no previous studies of these questions. Consequently, in this chapter, we provide a theoretical analysis of these hitherto unstudied questions.

The rest of this chapter is organized as follows: Section 2.1 describes our discrete-time Markov chain theoretic model[8] of an arbitrary renewable resource. Section 2.2 provides an answer to the how long until crisis question for a management regime that we call lax. Section 2.3 answers the same question for a management regime that we term strict. Section 2.4 uses a numerical example to illustrate the methods used in the previous two sections. Section 3 concludes and discusses possible extensions of this chapter's research.

---

6. See Lal (1991) and Dixon and Lal (1997) for discussions of the application of safe minimum standards in developing countries.

7. For more on short duration grazing, see Batabyal (2001; 2002), Holechek *et al.* (2001), and Batabyal and Yoo (2002).

8. Standard textbook accounts of discrete-time Markov chains can be found in Ross (1996; 2000) and in Taylor and Karlin (1998).

2. How Long Until Crisis?

2.1. *The Theoretical Framework*

A stochastic process $\{Z_t\}$ is said to be a Markov process if it possesses the property that, given the value of Z_t, the values of Z_u for $u > t$ are not affected by the values of Z_s for $s < t$. In words, the likelihood of any specific future behavior of the process, when its present state is known, is not changed by additional knowledge pertaining to its past behavior. A discrete-time Markov chain is a Markov process whose state space is either finite or countable and whose time index set is given by $I = (0, 1, 2, 3, \ldots)$. In symbols (see Taylor and Karlin, 1998, p. 95), the Markov property is

$$Pr\{Z_{n+1} = j / Z_0 = i_0, \ldots, Z_{n-1} = i_{n-1}, Z_n = i\} = Pr\{Z_{n+1} = j / Z_n = i\}, \tag{1}$$

for all time points n and all states $i_0, \ldots, i_{n-1}, i, j$. Further, the one-step transition probabilities of a discrete-time Markov chain are given by $P_{ij}^{n,n+1} = Pr\{Z_{n+1} = j / Z_n = i\}$ Finally, the reader should note that a discrete time Markov chain spends one time period in a state before making a transition to some state.

Consider an arbitrary renewable resource. Formally, we shall think of this resource as a discrete-time Markov chain with a finite number of rank ordered states $0, 1, 2, \ldots, S$. This means that the states toward the bottom of the rank order are undesirable states and the states toward the top of the rank order are the desirable states. In particular, state 0 is the least desirable (most undesirable) state and state S is the most desirable (least undesirable) state. Notionally, the reader might want to think of state S as one in which there is no human exploitation of the resource.

As indicated in Section 1, our resource will typically have a number of undesirable states but not all of them will be crisis states. We suppose that there is a single crisis state and we shall call this state, state 0. We now have to specify the state in which our manager inherits the resource under study. In principle, this could be any state i, where $i = 1, \ldots, S$. However, in what follows, to place the crisis issue in sharp focus, we shall analyze the situation that is as favorable as possible to our manager. This most favorable situation arises when the manager inherits the resource in state S. Now, given a particular management regime, we are interested in determining the expected value of T, where T is the time until our resource hits the crisis state 0. In technical parlance, we say that state 0 is the absorbing state and T is the time of absorption of our discrete-time Markov chain. In symbols, we are interested in determining c_i, where $c_i = E[T/Z_0 = i]$. Note that i in this conditional expectation expression can, in general, be any state from $1, 2, \ldots, S$. However, as explained previously, we shall conduct the analysis for the case in which the resource is initially in state S. This completes the discussion of our theoretical framework. We now discuss the "length of time until crisis" issue in the context of first, a lax management regime, and then, a strict management regime.

2.2. *The Lax Management Regime*

Let us first describe the lax management regime probabilistically. To this end, suppose that our resource is in state S in time period n. Then, in time period $n+1$, the position of the resource is uniformly distributed over the relevant states $0, 1, 2, \ldots, S-1$. As indicated in Section 2.1, the manager inherits the resource, and hence our analysis begins, in state S. This management regime is lax in the sense that even though the manager inherits the resource in the best possible state S, his or her managerial actions are superfluous and hence they are largely unsuccessful in keeping the resource away from the undesirable states in general and the crisis state in particular. Therefore, given that the resource is now in state S, it is just as likely that the resource will next be in state 0 as it is that it will next be in state $S-3$.

Examples of such lax management regimes include elephant and rhinoceros management regimes in countries like Zimbabwe in East Africa. As Milliken *et al.* (1993) and Dublin *et al.* (1995) have pointed out, in many East African nations, a variety of factors including the dramatic decline in law enforcement have resulted in the precipitous decline of elephant and rhinoceros populations. As a result, "black rhinos are locally extinct over substantial areas in Africa" (Brown and Layton, 2001, p. 33). Extinction corresponds to our crisis state. Further, note that East African elephant and rhinoceros management regimes have been lax in the extreme because they have allowed the resource to hit the most undesirable crisis state.

We now compute the expected length of time until crisis for our lax management regime. We will make use of the expression $c_i = E[T/Z_0 = i]$, for all the relevant states i. Now observe that because a discrete-time Markov chain spends one time period in each state before making a transition, it follows that the time T until crisis is always at least one period long. This tells us that

$$c_1 = 1. \tag{2}$$

Using the fact that $c_1 = 1$, we can tell that

$$c_2 = 1 + \frac{1}{2}c_1 \Rightarrow c_2 = 1 + \frac{1}{2}. \tag{3}$$

In similar fashion, we get

$$c_3 = 1 + \frac{1}{3}c_1 + \frac{1}{3}c_2 \Rightarrow c_3 = 1 + \frac{1}{3} + \left(\frac{1}{3}\right)\left(\frac{3}{2}\right) = 1 + \frac{1}{2} + \frac{1}{3}, \tag{4}$$

and continuing in this manner, we eventually get

$$c_S = 1 + \frac{1}{S}c_1 + \frac{1}{S}c_2 + \cdots + \frac{1}{S}c_{S-1}. \tag{5}$$

Now solving all the Equations (2)–(5) simultaneously, we conclude that

$$c_S = E[T/Z_0 = S] = 1 + \frac{1}{2} + \frac{1}{3} + \cdots + \frac{1}{S} = \sum_{i=1}^{i=S} \frac{1}{i}. \tag{6}$$

Equation (6) gives us the answer that we are after. In words, this equation tells us that for the lax management regime, the expected time until crisis is given by the summation of the reciprocal of the finite states, the summation being carried out from states 1 to S. For the asymptotic case, i.e., when $S \to \infty$, the term $\sum_{i=1}^{i=S}(1/i)$

is well approximated by the natural logarithm function. The use of this approximation gives us a more compact answer to the how long until crisis question. In particular, we get

$$c_S = E[T/Z_0 = S] = \log_e(S). \qquad (7)$$

In other words, the expected amount of time until our laxly managed resource hits the crisis state is given by the logarithm of the state in which the manager inherits the resource.

As noted in Holechek *et al.* (2001, p. 185), it is common for range managers to think of the condition of a rangeland in terms of four states called poor, fair, good, and excellent. Let us rank order these four states so that state 0 is the poor or crisis state and state 4 is the excellent state. Now, assuming that our lax manager inherits the rangeland in the excellent state, then, Equation (6) tells us that it will take, on average, 2.08 time periods for the resource to reach the crisis state.

Our analysis thus far permits us to draw a salient conclusion concerning the management of renewable resources in developing countries. Equations (6) and (7) clearly tell us that the answer to our how long until crisis question depends on state S, i.e., on the state in which our resource manager inherits the resource. Hence, from a management perspective, it is important to comprehend that the initial condition *matters*. More generally, given our rank ordering of the states of the resource, the more desirable the state in which the manager inherits the resource, the longer it will take for the resource to hit the crisis state. In contrast, if the initial state of inheritance is relatively undesirable, then it will take much less time for the resource to hit the crisis state. We now move on to answer the "how long until crisis" question when the resource under study is managed by a strict manager.

2.3. *The Strict Management Regime*

Once again, we begin with a probabilistic description of the strict management regime. Assume that our resource is in state j in time period n. Then, in time period $n+1$, the position of the resource is state 0 (the crisis state) with probability $1/j$, and its position is state k, where k is any one of the states $1, 2, 3, \ldots, j-1$, with probability $2k/j^2$. As in Section 2.2, our manager inherits the resource, and therefore our analysis commences, in state S. The present management regime is strict in two senses. First, given that the resource is now in state S, the likelihood of hitting the crisis state next is *not* identical to the likelihood of hitting some other (non-crisis) state. Second, the probability of hitting the crisis state next depends on where the resource is initially. In particular, the more desirable the initial state, i.e., the closer the initial state is (in rank order terms) to state S, the less likely it is that the resource will hit the crisis state.

There are many developing country management regimes that are strict in the two senses in which we are using the term strict in this chapter. Examples include the management of forest lands in the seven north-eastern states of India by shifting cultivators (see Agarwal and Narain (1985) and Noronha (1997)), and the management of offshore fishing in Bahia, Brazil (see Cordell and McKean (1986) and Noronha (1997)). In this Brazilian case, local fishermen have developed an intricate system of rules and enforcement mechanisms that permit "the fishermen to maintain a considerable jointness of use of the inshore fishery as a whole" (Cordell and McKean, 1986, p. 89). The reader will recall that whereas the dramatic decline in law enforcement was one of the contributory causes of the local extinction of black rhinos in East Africa (see Section 2.2), in the aforementioned Brazilian case, it is the existence of enforcement mechanisms that appear to have prevented the inshore fishery from hitting a crisis state.

Let us now calculate the average time until crisis for our strict management regime. Although we will once again use the fact that $c_i = E[T/Z_0 = i]$ for all the pertinent states i, our plan of action for computing c_S will be different. Using the logic employed in Section 2.2, it can be shown that our task now is to find a solution to the following equation in c_S

$$c_S = 1 + \sum_{j=1}^{j=S-1} \left(\frac{2j}{S^2}\right) c_j. \qquad (8)$$

To this end, we conjecture that

$$c_S = 2\left(\frac{S+1}{S}\right)\left(1 + \frac{1}{2} + \cdots + \frac{1}{S}\right) - 3 \quad (9)$$

solves Equation (8). To demonstrate that the expression in (9) does indeed solve Equation (8), let us change variables so that $C_k = kc_k$. With this change of variable, we now have to show that

$$C_k = 2(k+1)\left(1 + \frac{1}{2} + \cdots + \frac{1}{k}\right) - 3k \text{ solves} \quad (10)$$

$$C_S = S + \frac{2}{S}(C_1 + \cdots + C_{S-1}).$$

Using $C_k = kc_k$, we now use sums of numbers and we interchange order. This gives

$$\sum_{k=1}^{k=S-1} C_k = \sum_{k=1}^{k=S-1}\sum_{l=1}^{l=k} 2(k+1)\frac{1}{l} - 3\sum_{k=1}^{k=S-1} k. \quad (11)$$

Upon further simplification, the right-hand side (RHS) of Equation (11) equals

$$\sum_{l=1}^{l=S-1} \frac{2}{l} \sum_{k=l}^{k=S-1} (k+1) - \frac{3S(S-1)}{2}$$

$$= \sum_{l=1}^{l=S-1} \frac{2}{l}\left[\frac{S(S+1)}{2} - \frac{l(l+1)}{2}\right] - \frac{3S(S-1)}{2}. \quad (12)$$

Given Equations (11) and (12), it follows that

$$S + \frac{2}{S}\sum_{k=1}^{k=S-1} C_k = 2(S+1)\sum_{l=1}^{l=S-1}\frac{1}{l} - 3S + \frac{2(S+1)}{S} = C_S, \quad (13)$$

which is what we wanted to show (see Equation (10)).

In other words, the expression on the RHS of Equation (9), i.e., our conjectured solution, is indeed the solution that we are after. Specifically, Equation (9) tells us that for the strict management regime, the expected time until crisis is given by the product of two terms less the constant 3. The first term in this product is the ratio $2(S+1)/S$ and the second term is the summation of the reciprocal of the finite states, where the summation is carried out from states 1 to S. Comparing Equations (6) and (9), we see that this second term in the product on the RHS of Equation (9) is *identical* to the solution that we obtained to the how long until crisis question for the lax management regime. In other words, the two answers to the how long until crisis question are clearly related to each other.

Now, for the two management regimes that we have studied in this chapter, suppose that the number of states of the resource is the same, i.e., S, and that the manager inherits the resource in the same state, i.e., in state S. In this case, we would expect the time taken until crisis in the strict management regime to exceed the time taken until crisis in the lax management regime. A quick check with the four state range example of Section 2.2 tells us that this is indeed the case. In particular, it takes 2.08 time periods for the resource to hit the crisis state in the lax management regime, and — using Equation (9) we see that — it takes 2.21 time periods for the resource to hit the crisis state in the strict management regime.

What happens in the strict management regime when the number of states of the resource approaches infinity? Taking the limit as $S \to \infty$ in Equation (9), we get

$$c_S = E[T/Z_0 = S] = 2\log_e(S) - 3. \qquad (14)$$

In other words, the expected amount of time until our strictly managed resource hits the crisis state is given by twice the logarithm of the state in which the manager inherits the resource less the constant 3. Comparing the Equation (14) result with the corresponding result in Equation (7), we see that even in the asymptotic case, it generally takes longer for the resource to hit the crisis state with the strict management regime. As a quick check, when $S = 1,000,000$, using Equations (7) and (14) it is straightforward to verify that it takes approximately 14 time periods to hit the crisis state with the lax management regime and it takes about 25 time periods to hit the crisis state with the strict management regime.

As far as the management of renewable resources in developing countries is concerned, Equation (14) plainly tells us that the answer to our how long until crisis question, once again, depends on the state (S) in which the resource manager inherits the resource. Therefore, from a management perspective, as in Section 2.2, the initial condition *matters*. More generally, given our rank ordering of the states of the resource, if the initial state of inheritance is relatively undesirable, then it will take a small amount of time for the resource to hit the crisis state. In contrast, the more desirable the state in which the manager inherits the resource, the longer it will take for the resource to hit the crisis state. Keeping in mind the four state range example that we used in this and the previous section, we now provide a numerical example to show how the average time until crisis is computed when the transition probabilities of the discrete-time Markov chain are known.

2.4. *A Numerical Example*

Assume that the renewable resource under study is a rangeland with four states called Poor, Fair, Good, and Excellent. Reversing the rank ordering employed in the previous two sections, we shall say that state 0 is the Excellent state, state 1 is the Good state, state 2 is the Fair state, and state 3 is the Poor state. Further, we suppose that the transition probability matrix of this four state rangeland is

$$P = \begin{bmatrix} P_{00} & P_{01} & P_{02} & P_{03} \\ P_{10} & P_{11} & P_{12} & P_{13} \\ P_{20} & P_{21} & P_{22} & P_{23} \\ P_{30} & P_{31} & P_{32} & P_{33} \end{bmatrix} = \begin{bmatrix} 0.4 & 0.3 & 0.2 & 0.1 \\ 0 & 0.7 & 0.2 & 0.1 \\ 0 & 0 & 0.9 & 0.1 \\ 0 & 0 & 0 & 1 \end{bmatrix}. \quad (15)$$

Inspecting the above transition probability matrix, it is clear that states 0, 1, and 2 are not crisis states. However, because $P_{33} = 1$, state 3, the so-called "poor" state, is the only crisis state. Consistent with the discussion in the previous two sections of this chapter, suppose that our manager inherits this rangeland in state 0, the most desirable or "excellent" state. Now, given that the initial state is 0, the first row of the P matrix above is relevant. In particular, we see that as far as transitions are concerned, there are three possibilities. The rangeland may next hit the crisis state 3 in which case the expected time that we are after is 1 period long. However, the rangeland may stay in state 0 or it may next visit states 1 or 2. Therefore, to compute the expected time until crisis, we have to weight these various contingencies with the relevant probabilities that these contingencies will in fact occur. Doing so (see Taylor and Karlin, (1998, pp. 116–123) for the methodology), we see that the pertinent expected time that we seek is 10. In other words, in this example, the answer to the how long until crisis question is that it will take 10 time periods for the rangeland to hit the crisis state 3.

3. Conclusions

In the context of renewable resource management in developing countries, we provided a theoretical perspective on the hitherto unstudied "how long until crisis" question. In particular, we analyzed a lax and a strict management regime and we obtained answers to the expected time until crisis question for both the finite S and the large S cases. Further, we showed that the expected time until crisis is shorter in the lax management regime and longer in the strict management regime. Finally, we demonstrated that the answer to the how long until crisis question depends on the starting state. This is the reason for saying that the initial condition or history matters in renewable resource management.

The analysis of this chapter can be extended in a number of different directions. In what follows, we suggest two possible extensions. First, we investigated models in which there is a single crisis state. Consequently, an obvious way to generalize the analysis of this chapter would be to study the how long until crisis question with models in which there is more than one crisis state. Second, we used a discrete-time Markov chain to model and then analyze the lax and the strict management regimes. A drawback of this approach is that the time spent in a state before making a transition to some other state is deterministic and always one period long. The plausibility of this assumption is open to question. Therefore, it would be useful to eschew this assumption and research the how long until crisis question with more general models. Studies that incorporate these aspects of the problem into the analysis will increase our knowledge of the intricacies of renewable resource management in developing countries.

References

Agarwal, A. and Narain, S. (eds.) (1985). *The State of India's Environment 1984–85*. New Delhi, India: Center for Science and Environment.

Agarwal, B. (2000). Conceptualising Environmental Collective Action: Why Gender Matters. *Cambridge Journal of Economics* 24:283–310.

Batabyal, A.A. (2001). Some Theoretical Aspects of Short Duration Grazing. *International Journal of Ecology and Environmental Sciences* 27:215–220.

Batabyal, A.A. (2002). An Upper Bound for the Transient Resilience of a Rangeland that is Managed Using Cell Grazing. *Journal of Economic Research* 7:151–160.

Batabyal, A.A. and Yoo, S.J. (2002). The Steady State Distribution of Animals in Short Duration Grazing. *Applied Economics Letters* 9:695–698.

Batabyal, A.A. and Godfrey, E.B. (2002). Rangeland Management Under Uncertainty: A Conceptual Approach. *Journal of Range Management* 55:12–15.

Brown, G.M. and Layton, D.F. (2001). A Market Solution for Preserving Biodiversity: The Black Rhino. In J.F. Shogren and J. Tschirhart (eds.), *Protecting Endangered Species in the United States*. Cambridge, UK: Cambridge University Press.

Center for Science and Environment. (1990). *Human-Nature Interactions in a Central Himalayan Village: A Case Study of Village Bembru*. New Delhi, India: Center for Science and Environment.

Clark, C.W. (1990). *Mathematical Bioeconomics*, 2nd edn. New York, NY: Wiley.

Cordell, J.C. and McKean, M.A. (1986). Sea Tenure in Bahia, Brazil, in BOSTID. *Proceedings of the Conference on Common Property Resource Management*. Washington, DC: National Academy Press.

Dasgupta, P. (1996). The Economics of the Environment. *Environment and Development Economics* 1:387–428.

Dasgupta, P. and Maler, K.G. (1997). The Resource-Basis of Production and Consumption: An Economic Analysis. *In* P. Dasgupta and K.G. Maler (eds.), *The Environment and Emerging Development Issues*, Vol. 1. Oxford, UK: Clarendon Press, Oxford.

Dixon, J. and Lal, P. (1997). The Management of Coastal Wetlands: Economic Analysis of Combined Ecologic-Economic Systems. *In* P. Dasgupta and K.G. Maler (eds.), *The Environment and Emerging Development Issues*, Vol. 1. Oxford, UK: Clarendon Press, Oxford.

Dublin, H.T., Milliken, T. and Barnes, R.F.W. (1995). *Four Years After the CITES Ban*. Report of IUCN/SSC African Elephant Specialist Group. Cambridge, UK: Traffic International.

Falconer, J. (1990). *The Major Significance of Minor Forest Products*. Rome, Italy: Food and Agriculture Organization.

Falconer, J. and Arnold, J.E.M. (1989). *Household Food Security and Forestry*. Rome, Italy: Food and Agriculture Organization.

Ho, P. (2000). China's Rangelands Under Stress: A Comparative Study of the Pasture Commons in the Ninxia Hui Autonomous Region. *Development and Change* 31:385–412.

Holechek, J.L., Pieper, R.D. and Herbel, C.H. (2001). *Range Management*, 4th edn. Upper Saddle River, NJ: Prentice Hall.

Jodha, N.S. (1980). The Process of Desertification and the Choice of Interventions. *Economic and Political Weekly* 15:1351–1356.

Lal, P. (1991). Utilization and Management of Coastal Wetland Resources in Kosrae, in Pacific Island Network. *Kosrae Island Resource Management Plan*. Manoa, HI: University of Hawaii.

Milliken, T., Nowell, K. and Thomsen, J.B. (1993). *The Decline of the Black Rhino in Zimbabwe*. Cambridge, UK: Traffic International.

Noronha, R. (1997). Common-Property Resource-Management in Traditional Societies, *In* P. Dasgupta and K.G. Maler (eds.), *The Environment and Emerging Development Issues*, Volume 1. Oxford, UK: Clarendon Press, Oxford.

Ross, S.M. (1996). *Stochastic Processes*, 2nd edn. New York: Wiley.

Ross, S.M. (2000). *Introduction to Probability Models*, 7th edn. San Diego, CA: Harcourt Academic Press.

Taylor, H.M. and Karlin, S. (1998). *An Introduction to Stochastic Modeling*, 3rd edn. San Diego, CA: Academic Press.

Wade, R. (1988). *Village Republics*. Cambridge, UK: Cambridge University Press.

Chapter 7

A STACKELBERG GAME MODEL OF TRADE IN RENEWABLE RESOURCES WITH COMPETITIVE SELLERS

With Hamid Beladi

We model international trade in renewable resources between a single buyer and competitive sellers as a Stackelberg differential game. The buyer uses unit and *ad valorem* tariffs to indirectly encourage conservation of the renewable resource under study. First, we show that the efficacy of these trade policy instruments in promoting conservation depends fundamentally on whether harvesting costs are stock dependent or independent. When harvesting costs are stock independent, the optimal open loop tariffs are dynamically consistent. In contrast, when harvesting costs are stock dependent, the optimal open loop tariffs are dynamically inconsistent. Second, we point out that whether the terminal value of the resource stock is higher with the stock independent or the stock dependent cost function cannot be resolved unambiguously. Third, we show that it does not make sense for the buyer to use both tariffs simultaneously. Finally, we discuss the implications of these and other findings for renewable resource conservation in general.

1. Introduction

Renewable resources such as fish, timber (from forests), ivory (from elephants) and horns (from rhinoceroses) have been traded between countries for quite some time. However, considerable concern has recently been expressed about the declining stock levels of most salient renewable resources.[1] As such, researchers and

---

1. For more on this, see Clark (1973), Jablonski (1991), and Pimm *et al.* (1995).

conservationists have now begun to systematically analyze questions about (i) the desirability of international trade in renewable resources, and (ii) the efficacy of trade policies in promoting the conservation of these renewable resources.

This chapter is concerned with the second of these two questions. However, before we move to the specifics of the chapter itself, let us first summarize the relevant literature on international trade in renewable resources. Barbier and Schulz (1997, pp. 160–161) conclude that "ambiguous stock effects make trade interventions a poor policy instrument for securing biodiversity conservation." Schulz (1997) shows that the effects of trade sanctions depend not only on the bioeconomic interactions between the species but also on the management system in the targeted country. Consequently, the threat of trade sanctions will not necessarily result in lower harvesting and higher stocks of marine mammals.

In an interesting two-country, two-good model of trade in renewable resources, Brander and Taylor (1998) demonstrate that not only is the basic "gains from trade" idea undermined by the presence of open access renewable resources but that tariffs imposed by a resource importing country always benefit the resource exporter. Examining a two country model of the effects of unilateral fishery management, Emami and Johnston (2000) argue that the trade induced losses that arise from not managing this fishery can be mitigated by levying import tariffs on the resource good. Maestad (2001) shows that depending on the manner in which trade restrictions affect the log prices of alternate tree qualities, trade restrictions may decrease or increase timber logging.

The question of the efficacy of trade policies in promoting the conservation of renewable resources has probably been discussed most extensively in the context of the ivory trade between developing countries in Africa and East Asian and western nations. In an early contribution, Barbier *et al.* (1990) argued against a ban on trade in ivory and proposed an alternate strategy. This strategy would permit limited trade in ivory and the objective of this strategy would be to create sufficient incentives for the sustainable management of African elephant populations. In the aftermath of the CITES ban on ivory trade, Bulte and van Kooten (1999) examined the desirability of permitting some ivory trade. Bulte and van Kooten (1999) caution against lifting the trade ban. Specifically, they point out that permitting some trade in ivory would encourage illegal poaching and that this could drive the African elephant to extinction. In a more recent contribution, Heltberg (2001) uses a numerical model to conclude that the ivory trade ban is likely to reduce poaching.

Although these studies have certainly advanced our understanding of many aspects of international trade in renewable resources, none of these studies have analyzed the connections between renewable resource harvesting costs and trade policy. In particular, how does the stock dependence/independence of harvesting costs affect the efficacy of trade policy? Although researchers thus far have not studied this question, as we shall see, the form of the harvesting cost function has profound implications for the efficacy of trade policy in promoting the conservation of renewable resources.

The rest of this chapter is organized as follows. Section 2 provides a detailed description of the Stackelberg differential game model of international trade in a renewable resource between a single buyer and competitive sellers. Section 3 analyzes the effects of trade policy when the renewable resource harvesting cost function is stock independent. Section 4 does the same for the case in which the harvesting cost function is stock dependent. Finally, Section 5 concludes and offers suggestions for future research on international trade in renewable resources.

2. The Stackelberg Differential Game

Our model is adapted from Karp (1984) and Batabyal (1996). There is a single buyer (the leader) of the renewable resource and this buyer faces competitive sellers (the followers). If the renewable resource is ivory, then the reader should think of our model as a description of the interaction between competitive ivory sellers in the developing countries of Africa and a buyer with market power such as Japan.[2] Denote the stock of the renewable resource at time t by $x(t)$.[3] The buyer's utility from consuming the resource at harvest level $h(t)$ is given by the concave and differentiable utility function $u(h)$. The domestic market in the buyer's country is competitive so that $u'(h) = p(h)$, the price that consumers in the importing nation pay for this resource. The government of the importing nation has access to two trade policy instruments: a unit tariff denoted by $n(t)$ and an *ad valorem* tariff denoted by $a(t)$, where $v = 1/(1 + a)$. When the government in the importing nation uses the unit tariff $n(t)$, the price received by the competitive sellers (exporters) is $p(h) - n(t)$. Similarly, when this government uses the *ad valorem* tariff $a(t)$, the price received by the exporters is $v(t)p(h)$.

The buyer's payoff in the finite horizon (from time $t = 0$ to $t = T$) games to be analyzed in this chapter is the discounted stream of the difference between the utility of consuming the resource at level $h(t)$ and the payment to the competitive sellers. Consequently, if we denote the interest rate by r, then the buyer's payoff is

$$J_b = \int_0^T e^{-rt}[u(h) - \{p(h) - n\}h]dt, \qquad (1)$$

when he uses a unit tariff. Similarly, when he uses the *ad valorem* tariff, his payoff is

$$J_b = \int_0^T e^{-rt}[u(h) - vp(h)h]dt. \qquad (2)$$

The competitive seller maximizes profit and the seller gets no utility from consuming the resource under study. A major objective of this chapter is to demonstrate the dependence of optimal trade policy on the cost of harvesting the renewable resource x. To this end, we analyze two kinds of harvesting cost functions. The first

---

2. For a splendid account of the African elephant in general and the ivory trade in particular, see Meredith (2001).

3. In the rest of this chapter we shall frequently suppress the time dependence of the various variables. However, the reader should note that all the variables that we work with depend on time.

kind of cost function — analyzed in Section 3 — is stock *independent* and it depends only on the harvest level. Denote this thrice differentiable cost function by $c(h)$, where h is harvest, $c'(h) \geq 0$ and $c''(h) \geq 0$. The second kind of cost function — analyzed in Section 4 — depends on the stock and on the harvest level. Let us denote this cost function by $c(x)h$, where $c'(x) \leq 0$ and $c''(x) \geq 0$. *Ceteris paribus*, the stock-independent cost function is more relevant when the amount harvested is small relative to the total size of the resource stock. In contrast, the stock dependent cost function is more relevant when the total size of the stock is small to begin with and/or when the amount harvested is a significant fraction of the total stock size. When the buyer uses a unit tariff, the seller's payoff is

$$J_s = \int_0^T e^{-rt}[p(h)h - nh - C(\cdot)]dt \qquad (3)$$

and when this buyer uses the *ad valorem* tariff, the seller's payoff is

$$J_s = \int_0^T e^{-rt}[vp(h)h - C(\cdot)]dt, \qquad (4)$$

where $C(\cdot)$ in Equations (3) and (4) is $c(h)$ and $c(x)h$, respectively.

The buyer controls the tariffs $n(t)$ and $v(t)$, and the seller controls the harvest $h(t)$. As the leader, the buyer announces a tariff trajectory at the beginning of the game and the seller takes this trajectory as given. The buyer and the seller are constrained by the differential equation describing the dynamics of the resource stock. That equation is

$$dx(t)/dt = \dot{x} = bx - h(t), \qquad (5)$$

where $0 < b < r < \infty$ and $x(0) = x_0 > 0$ is given. In other words, the temporal net change in the resource stock is the difference between the natural growth bx and the harvest $h(t)$.[4]

---

4. Replacing the linear natural growth function with a general growth function $f(x)$ does not alter the main points that we wish to make in this chapter. However, the algebra associated with the various derivations becomes significantly more complicated. This is the reason for sticking with the linear growth function.

The terminal value of the renewable resource stock depends in part on the tariff used by the buyer. One can think of these alternate terminal stock levels as the outcomes of different games. One way of comparing these outcomes is to compare the resultant stock levels. In this connection, we say that game 1 is more resource conserving than game 2 if and only if $x_1^T > x_2^T$, where $x_i^T, i = 1, 2$ is the terminal value of the resource stock in game $i = 1, 2$. We determine the different harvest trajectories by deriving a differential equation satisfied by the equilibrium $h(t)$. When we are able to compare the various differential equations without resorting to additional assumptions, we shall do so. However, the reader should note that because of the complexity of the underlying mathematics, in some cases it will not be possible to obtain general results.

Let us now consider the benchmark case in which there is free trade. In this case, the buyer is passive, he sets $n(t) = 0$ or $v(t) = 1$, and the seller solves a standard control problem. When the harvest cost function is stock independent, the optimal competitive harvest rate solves

$$\{p'(h) - c''(h)\}\dot{h} - (r - b)\{p(h) - c'(h)\} = 0, \qquad (6)$$

and when the harvest cost function is stock dependent, the optimal competitive harvest rate solves

$$p'(h)\dot{h} - (r - b)\{p(h) - c(x)\} - bxc'(x) = 0. \qquad (7)$$

In what follows, we compare the differential equation satisfied by the optimal harvest level in Section 3 with Equation (6) and that in Section 4 with Equation (7).

3. The Stock Independent Harvest Cost Function and Open Loop Tariffs

We now derive the optimal open loop unit and *ad valorem* tariffs for our single buyer (monoposonist). The reader should note that although open loop tariffs are generally dynamically inconsistent (Karp and Newbery, 1993), for the case studied in this section, these open loop tariffs *are* dynamically consistent. Before we explain why this is the case, let us first understand what would happen were these tariffs to be dynamically inconsistent. If these tariffs were dynamically inconsistent, then at some time $t > 0$ the buyer would want — if he could — to deviate from the tariff trajectory he announced at the beginning of the game (at time $t = 0$) and announce a different tariff trajectory. The competitive seller in this chapter is forward looking. As such, she would anticipate the buyer's desire to change the tariff trajectory he announced at the beginning of the game and hence this tariff would fail to achieve its intended objectives. In the class of Stackelberg games analyzed in this chapter, the stock independence of the harvest cost function accounts for the dynamic consistency of the optimal solution. To see this clearly, we now derive, in turn, the open loop unit and the *ad valorem* tariffs.

3.1. *The Open Loop Unit Tariff*

We solve the buyer's problem using a method used previously by Simaan and Cruz (1973), Karp (1984), and Batabyal (1996). The basic idea is as follows: The buyer treat's the seller's first-order condition as an ordinary constraint and her costate variable as a state variable. These two conditions and suitable boundary conditions convert the differential game into a control problem for the buyer.

When the seller takes the buyer's unit tariff $n(t)$ as given, the first-order necessary conditions to her problem are

$$p(h) - n - c'(h) - \lambda = 0 \tag{8}$$

and

$$\dot{\lambda} = (r - b)\lambda, \tag{9}$$

where $\lambda(t)$ is the costate variable.[5] This costate variable gives us the seller's marginal utility of one more unit of the resource stock at

---

5. Equation (9) describes a jump state constraint. In other words, the initial value of λ, $\lambda(0)$, is free and the value of this jump state variable at any arbitrary point in time is determined by current and/or future events. Put differently, Equation (9) is not a fixed initial state constraint for the buyer. For more on jump state constraints see Karp and Newbery (1993).

time t. Now solving for n from Equation (8) and substituting in Equation (1), we get

$$J_b = \int_0^T e^{-rt}[u(h) - \{c'(h) + \lambda\}h]dt. \qquad (10)$$

Equation (10) gives the buyer's payoff as the present discounted stream of the difference between the utility of consuming $h(t)$ and the sum of the marginal harvest cost times the harvest $h(t)$ and the term λh. This last term is the total instantaneous rent paid by the buyer for the resource x.

In order to keep the buyer's problem a one state variable problem, let us eliminate λ from (10) by using (9). Solving Equation (9), it is clear that $\lambda(t) = 0$. Substituting this value of 8 into (10) we get

$$J_b = \int_0^T e^{-rt}[u(h) - c'(h)h]dt. \qquad (11)$$

Note that we have now converted the buyer's problem from one of maximizing (1) over n subject to (5) to one of maximizing (11) over h subject to (5). The first-order necessary conditions to this problem are

$$p(h) - c'(h) - hc''(h) - \sigma = 0, \qquad (12)$$

and

$$\dot{\sigma} = (r - b)\sigma, \qquad (13)$$

where $\sigma(t)$ is the costate variable. Inspecting (12), we see that the solution to the buyer's problem does *not* depend on the initial stock of the resource x_0. This explains why the optimal solution to the buyer's problem is dynamically consistent. Put differently, because it is optimal to set $\lambda(t) = 0$, the total instantaneous rent paid by the buyer for the resource x does not affect his maximization problem. Therefore, the question of altering the total instantaneous rent paid to the seller over time does not arise.

To find the differential equation satisfied by the optimal $h(t)$ when our buyer uses a unit tariff, differentiate (12) with respect to time and then use (13) to simplify the resulting expression. This yields

$$\{p'(h) - 2c''(h) - hc'''(h)\}\dot{h} - (r - b)\{p(h) - c'(h) - hc''(h)\} = 0, \qquad (14)$$

where $p\{h(T)\} = c'\{h(T)\} + h(T)c''\{h(T)\}$ is the boundary condition for h. Comparing Equations (14) and (6) it is clear that the optimal harvest level when the buyer uses a unit tariff is not the same as the optimal harvest level when this buyer is passive. When the harvest cost function is quadratic, i.e., when $c(h) = h^2/2$, the differential equation for the optimal harvest with a unit tariff approximates the differential equation for the optimal harvest when the buyer is passive. This notwithstanding, there are no straightforward necessary or sufficient conditions under which (6) and (14) coincide. The differential equation for the optimal unit tariff is found by differentiating (8) and then using (9) to simplify the ensuing expression. This gives

$$\dot{n} - (r - b)n = \{p'(h) - c'(h)\}\dot{h} - (r - b)\{p(h) - c'(h)\}, \qquad (15)$$

where $\dot{h}$ is given by (14) and $n(T) = p\{h(T)\} - c'\{h(T)\}$ is the boundary condition.

3.2. The Open Loop Ad Valorem Tariff

We now derive the solution for the open loop *ad valorem* tariff when the buyer's objective is given by (2). To maximize (4) subject to (5), we first form the seller's current value Hamiltonian. The pertinent first-order necessary conditions are

$$vp(h) - c'(h) - \lambda = 0, \qquad (16)$$

and (9). Now solve for v from (16), substitute into (2), and then simplify the resulting expression. We get

$$J_b = \int_0^T e^{-rt}[u(h) - \{c'(h) + \lambda\}h]dt. \qquad (17)$$

Comparing Equations (17) and (10) it is clear that these two objective functionals are identical. This tells us that the renewable resource buyer's payoff is policy *invariant*. Further, note that because $\lambda(t) = 0$, the constraint in both cases is (5). From this we conclude that the open loop unit and *ad valorem* tariffs are equivalent. Consequently, they will both give rise to the same time profile of harvests and hence to the same terminal level of the resource stock. We now discuss the implications of these findings.

3.3. *Discussion*

From the previous paragraph, we conclude that the unit payment from the buyer to the seller is $p(h) - n(t)$ in the case of the unit tariff and $v(t)p(h)$ in the case of the *ad valorem* tariff. Moreover, because $\lambda(t) = 0$, it is easy to verify that $p(h) - n(t) = v(t)p(h) = c'(h)$. Put differently, when a single buyer trades with competitive sellers, this buyer gains nothing by using both tariffs simultaneously. Indeed, one tariff is superfluous and it is optimal to set this tariff equal to zero.

Equation (13) tells us that $\sigma(t) = 0$, $\forall t$. In other words, the buyer's marginal utility of one additional unit of the resource stock is zero. Using this in (12) we get an equation for the domestic price of the renewable resource in the importing nation. That equation is

$$p(h) = c'(h) + h(t)c''(h), \quad \forall t. \tag{18}$$

The reader will note that the domestic price of the renewable resource in the importing nation does not depend on the initial value of the resource stock x_0.

Since the optimal unit and *ad valorem* tariffs are equivalent, a comparative exercise involving these two trade policy instruments is not relevant. Recall that the objective of the importing nation is to encourage the conservation of the renewable resource. In this regard we note that because the tariffs studied here are dynamically consistent, they will achieve their intended conservation objectives, albeit indirectly. Having said this, we should also point out that if an importing nation's objective is to encourage conservation of the renewable resource in the exporting countries, then tariffs are not the ideal policy instruments. Why not? This is because tariffs target trade and the direct effect of the tariff is to discourage domestic consumption in the importing nation. Tariffs do not do anything directly to encourage conservation of the renewable resource in the exporting nations. Consequently, from the standpoint of resource conservation, tariffs are blunt policy instruments. We now analyze the Stackelberg differential game between our single buyer and competitive sellers when the harvest cost function is stock dependent.

4. The Stock Dependent Harvest Cost Function and Open Loop Tariffs

When the harvest cost function is stock dependent, the optimal open loop unit and *ad valorem* tariffs are dynamically inconsistent. As indicated in Section 3, this means that at some time $t > 0$ the buyer will want — if he can — to deviate from the tariff trajectory he announced at time $t = 0$ and announced a different tariff trajectory. The competitive seller in this chapter is forward looking. Therefore, she will foresee the buyer's desire to alter the tariff trajectory he announced at the beginning of the game and hence this tariff will fail to attain its intended conservation goals. To comprehend this crucial feature of the optimal policies clearly, we now derive the optimal open loop unit and *ad valorem* tariffs.

4.1. *The Open Loop Unit Tariff*

Recall from Section 2 that the stock dependent harvest cost function is $c(x)h$. We follow the Section 3 method to solve our buyer's problem. Suppose the seller takes the buyer's unit tariff $n(t)$ as given. Then the first-order necessary conditions to her problem are

$$p(h) - n - c(x) - \lambda = 0 \qquad (19)$$

and

$$\dot{\lambda} = (r - b)\lambda + c'(x)h, \qquad (20)$$

where $\lambda(t)$ is the costate variable. This costate variable gives us the seller's marginal utility of one more unit of the resource stock at time t. Comparing (20) with (9) it is immediately clear that when the harvest cost function is stock dependent, $\lambda(t) \neq 0$. In other words, the rent on the marginal unit of the resource stock is typically not equal to zero. Now solving for n from Equation (19) and then substituting into Equation (1), we get

$$J_b = \int_0^T e^{-rt}[u(h) - \{c(x) + \lambda\}h]dt. \tag{21}$$

Equation (21) gives the buyer's payoff as the present discounted stream of the difference between the utility of consuming $h(t)$ and the sum of the cost of harvesting $h(t)$ and the term λh. As in Section 3, this last term can be interpreted as the total instantaneous rent paid by the buyer for the resource x.

In order to keep the buyer's problem a single state variable problem, we now eliminate λ from (21) by using (20). Integrating Equation (20) we get

$$\lambda(t) = e^{(b-r)(T-t)}\lambda(T) - e^{-(b-r)t}\int_t^T e^{(b-r)m}c'(x)hdm. \tag{22}$$

Substituting this value of $\lambda(t)$ from (22) into (21) we obtain

$$J_b = \int_0^T e^{-rt}[u(h) - c(x)h - \{e^{(b-r)(T-t)}\lambda(T) - e^{-(b-r)t}$$
$$\times \int_t^T e^{(b-r)m}c'(x)hdm\}h]dt. \tag{23}$$

Equation (23) tells us that for any time path of $h(t)$, the buyer's objective functional is maximized by setting $\lambda(T) = 0$. So, the buyer drives the seller's rent to zero at the end of the game. Now substitute $\lambda(T) = 0$ into (23) and reverse the order of integration of the last term in (23). This gives

$$J_b = \int_0^T e^{-rt}[u(h) - c(x)h + c'(x)h\{e^{bt}x_0 - x\}]dt. \tag{24}$$

As desired, we now have a single state variable problem for our buyer. Specifically, we have converted the buyer's problem from one of maximizing (1) over n subject to (5) to one of maximizing (24) over h subject to (5). Comparing Equations (24) and (11) we see that unlike the case analyzed in Section 3, when the harvest cost function is stock dependent, the initial value of the resource stock x_0 enters the buyer's objective functional. The first-order necessary conditions to our buyer's problem are

$$p(h) - c(x) + c'(x)\{e^{bt}x_0 - x(t)\} - \sigma = 0 \qquad (25)$$

and

$$\dot{\sigma} = (r - b)\sigma + 2c'(x)h - c''(x)h\{e^{bt}x_0 - x(t)\}, \qquad (26)$$

where $\sigma(t)$ is the costate variable. Equation (25) tells us that when the harvest cost function is stock dependent, the solution to the buyer's problem does depend on the initial stock of the renewable resource x_0. What this means is that if the buyer were able to revise — at some time $m > 0$ — the tariff he initially announced at the beginning of the game, then x_0 in (25) would have to be replaced with $x(m)$ and the solution for all $t > m$ would be different. In other words, the buyer's optimal solution is dynamically inconsistent. From (25) and (26) it is clear that the optimal solution is dynamically consistent if and only if the harvest cost function is unrelated to the stock of the resource. Indeed, this is what we demonstrated in our analysis of the stock independent harvest cost function in Section 3.

Before deriving a differential equation satisfied by the optimal $h(t)$, a comment on the salience of constraint (20) is in order. The seller of the renewable resource in this chapter has no market power and she solves a dynamic but standard control problem. In contrast, the buyer exerts market power and he is constrained by the dynamic maximizing behavior of the seller. These two features of the buyer's problem together tell us that his control problem is non-standard. Put differently, because the seller's problem is dynamic and the buyer exerts market power, a "rational expectations" constraint — given by Equation (20) — is introduced into the buyer's problem.[6] The presence of this constraint makes our buyer's problem a non-standard control problem.

To find the differential equation satisfied by the optimal $h(t)$ when our buyer uses a unit tariff, we differentiate (25) with respect to time and then use (26) to simplify the resulting expression. This process gives

$$p'(h)\dot{h} - (r-b)\{p(h) - c(x) + e^{bt}x_0 c'(x) - c'(x)x\}$$
$$+ bc'(x)\{e^{bt}x_0 - 2x\} + c''(x)\{bx_0 x - hx_0 - bx^2 + e^{bt}hx_0\} = 0, \quad (27)$$

where $p\{h(T)\} = c\{x(T)\} - c'\{x(T)\}\{e^{bT}x_0 - x(T)\} + \sigma(T)$ is the boundary condition for h. Inspecting (27), it is clear that the exogenously given initial condition $x(0) = x_0$ affects the temporal behavior of the optimal harvest $h(t)$. Comparing Equations (27) and (7) we see that as in Section 3, the optimal harvest level when our buyer uses a unit tariff and the harvest cost function is stock dependent is not the same as the optimal harvest level when this buyer is passive. Moreover, there are no obvious necessary or sufficient conditions under which Equations (7) and (27) coincide. The differential equation for the optimal unit tariff is found by differentiating (19) with respect to time and then using (20) to simplify the resulting expression. This gives

$$\dot{n} - (r-b)n = (r-b)\{e^{bt}x_0 c'(x) - c'(x)x\} - bc'(x)\{e^{bt}x_0 - x\}$$
$$- c''(x)\{bx_0 x - hx_0 - bx^2 + e^{bt}hx_0\}, \quad (28)$$

where $\dot{h}$ is given by (27) and the boundary condition is $n(T) = p\{h(T)\} - c\{x(T)\}$.

4.2. *The Open Loop Ad Valorem Tariff*

Even when the harvest cost function is stock dependent, the optimal open loop unit and *ad valorem* tariffs are equivalent. We now

---

6. In Footnote 5, we called this "rational expectations" constraint a jump state constraint. For more on these matters, see Karp (1984) and Karp and Newbery (1993).

demonstrate this fact. When the seller takes the buyer's *ad valorem* tariff as given, the first-order necessary conditions to her control problem are

$$vp(h) - c(x) - \lambda = 0 \qquad (29)$$

and (20). Solving for v from (29), substituting into (2), and then simplifying, we get

$$J_b = \int_0^T e^{-rt}[u(h) - \{c(x) + \lambda\}h]dt. \qquad (30)$$

Comparing (30) and (21) it is clear that these two objectives' functionals are identical. Further, before the rent $\lambda(t)$ has been eliminated from the objective functionals (21) and (30), the constraints for both the problems are given by (5) and (20). This tells us that the optimal unit and the *ad valorem* tariffs are equivalent. In particular, both tariffs are plagued by the problem of dynamic inconsistency and it does not matter which tariff the buyer uses because his payoff with either tariff is the same. From (19) and (29), the reader can easily verify that $p(h) - n(t) = v(t)p(h) = c(x) + \lambda$. Hence, one tariff is redundant and our buyer gains nothing by using both tariffs in unison. We now discuss the nexuses between the form of the harvest cost function, the buyer's trade (tariff) policies, and the efficacy of these policies in promoting the conservation of the renewable resource under study.

4.3. *Discussion*

Upon rearranging terms in Equation (25), we get an equation for the domestic price of the renewable resource in the importing nation. That equation is

$$p(h) = c(x) - c'(x)\{e^{bt}x_0 - x(t)\} + \sigma(t), \quad \forall t, \qquad (31)$$

where $\sigma(t)$ is the buyer's marginal utility of an additional unit of the resource at time t. Note that because the harvest cost function is stock dependent, unlike the case examined in Section 3.3 (see Equation (18)), the buyer's marginal utility of one more unit of the resource stock $\sigma(t)$ is now no longer equal to zero. Also, the initial value of the resource stock now affects the domestic price of the renewable resource in the importing nation. Comparing Equations (18) and (31) we see that, in general, it is not possible to determine whether the domestic price is higher with the stock independent or with the stock dependent harvest cost function. Further, note that x_0 affects the optimal solution when the harvest cost function is stock dependent. In contrast, x_0 does not affect the optimal solution in the case of the stock independent harvest cost function. Consequently, in general, it is not possible to determine whether the terminal value of the resource stock $x(T)$ is higher with the stock independent or with the stock dependent harvest cost function.

The analysis in this chapter has shown that attempts to promote renewable resource conservation by means of trade policies are problematic in more ways than one. Let us now discuss this salient point in greater detail. The analysis in Karp and Newbery (1993) and in Batabyal (1998) tells us that given a choice between dynamically inconsistent and consistent policies, an economic agent will typically choose dynamically inconsistent policies because inconsistent policies result in a higher payoff. In the context of this chapter, this tells us that even when the open loop unit and *ad valorem* tariffs are dynamically inconsistent — and this happens when the harvest cost function is stock dependent — the buyer in the importing nation will prefer to use these inconsistent trade policies rather than follow a dynamically consistent course of action. However, inconsistent policies are not credible and hence the tariff trajectory announced by the buyer at the beginning of the game will not be believed by the sellers and therefore inconsistent policies will fail to achieve their resource conservation objectives.

In contrast, when the harvest cost function is stock independent, the optimal open loop tariffs are dynamically consistent and hence believable by the sellers of the resource. Consequently, in this case, the buyer's trade policies (tariffs) will, albeit indirectly, attain their resource conservation objectives. As indicated previously in Section 3.3, in an ideal situation, trade policies such as tariffs should not be used to promote the conservation of renewable resources. This is because tariffs reduce the domestic consumption of the traded resource in the importing nation and hence get at the conservation issue only indirectly. However, if sellers are unwilling and/or unable to take measures in their own countries to promote resource conservation, possibly because of the lack of apposite property rights, then tariffs are one imperfect instrument with which sellers can be encouraged to undertake the appropriate conservation measures.

Although tariffs are an imperfect way of promoting conservation, the analysis in this chapter has shown that they may not work as desired. This is because the credibility of the optimal tariffs depends on the form of the harvest cost function and this form is *not* controllable by the buyer. This crucial point has not been recognized previously in the literature on international trade in renewable resources. We have already noted that the stock dependent cost function is the more appropriate cost function for endangered renewable resources such as the African elephant. Resources like the African elephant are endangered at least in part because adequate domestic measures have not been taken in the pertinent countries to prevent overexploitation. In the case of the African elephant, the relevant countries include South Africa, Uganda, Zambia, and Zimbabwe. It is for these endangered resources — where the appropriate domestic conservation measures have not been taken — that imperfect supra-national measures such as trade policies are most needed. Unfortunately, the analysis in this chapter tells us that trade policies are likely to be ineffective (because they are not credible) precisely when they are most needed (when the harvest cost function is stock dependent).

5. Conclusions

This chapter studied a Stackelberg differential game model of international trade in renewable resources. We analyzed trade in a renewable resource between a single buyer who leads and competitive sellers who follow. The government in the importing nation uses trade policies (unit and *ad valorem* tariffs) to indirectly promote the conservation of the renewable resource being studied. The optimal open loop unit and *ad valorem* tariffs are equivalent and hence the buyer gains nothing by using both tariffs simultaneously. The efficacy of trade policies in promoting resource conservation depends fundamentally on the form of the harvest cost function. The analysis conducted here shows that because of credibility problems, trade policies such as tariffs are likely to be ineffective in promoting the conservation of threatened renewable resources.

The analysis in this chapter can be extended in a number of different directions. In what follows, we suggest two possible extensions. First, it would be useful to determine the properties of the optimal open loop tariffs when in addition to the buyer, the seller also obtains utility from consuming the renewable resource. Second, it would be interesting to study the trade game between our monopsonistic buyer and a monopolistic seller of the renewable resource. In this case the buyer and the seller would have market power. Consequently, if the buyer leads, then it would be interesting to know, *inter alia*, whether he is able to use trade policies to reduce or neutralize the seller's market power. An examination of these aspects of the problem will allow richer analyses of the connections between trade, trade policies, and the conservation of renewable resources.

References

Barbier, E.B., Burgess, J., Swanson, T. and Pearce, D. (1990). *Elephants, Economics, and Ivory*. London, UK: Earthscan.

Barbier, E.B. and Schulz, C.-E. (1997). Wildlife, Biodiversity, and Trade. *Environment and Development Economics* 2:145–172.

Batabyal, A.A. (1996). Consistency and Optimality in a Dynamic Game of Pollution Control I: Competition. *Environmental and Resource Economics* 8:205–220.

Batabyal, A.A. (1998). Environmental Policy in Developing Countries: A Dynamic Analysis. *Review of Development Economics* 2:293–304.

Brander, J. and Taylor, S. (1998). Open Access Renewable Resources: Trade and Trade Policy in a Two-Country Model. *Journal of International Economics* 44:181–209.

Bulte, E. and van Kooten, C. (1999). Economic Efficiency, Resource Conservation, and the Ivory Trade Ban. *Ecological Economics* 28:171–181.

Clark, C.W. (1973). The Economics of Overexploitation. *Science* 181: 630–634.

Emami, A. and Johnston, R. (2000). Unilateral Resource Management in a Two-Country General Equilibrium Model of Trade in a Renewable Fishery Resource. *American Journal of Agricultural Economics* 82:161–172.

Heltberg, R. (2001). Impact of the Ivory Trade Ban on Poaching Incentives: A Numerical Example. *Ecological Economics* 36:189–195.

Jablonski, D. (1991). Extinctions: A Paleontological Perspective. *Science* 253:754–757.

Karp, L.S. (1984). Optimality and Consistency in a Differential Game with Non-Renewable Resources. *Journal of Economic Dynamics and Control* 8:73–97.

Karp, L.S. and Newbery, D.M. (1993). Intertemporal Consistency Issues in Depletable Resources. *In* A. Kneese and J. Sweeny (eds.), *Handbook of Natural Resource and Energy Economics*, Vol. 3. Amsterdam, The Netherlands: Elsevier.

Maestad, O. (2001). Timber Trade Restrictions and Tropical Deforestation: A Forest Mining Approach. *Resource and Energy Economics* 23:111–132.

Meredith, M. (2001). *Africa's Elephant*. London, UK: Hodder and Stoughton.

Pimm, S., Russell, G. and Gittleman, J. (1995). The Future of Biodiversity. *Science* 269:347–350.

Schulz, C.-E. (1997). Trade Sanctions and Effects on Long-Run Stocks of Marine Mammals. *Marine Resource Economics* 12:159–178.

Simaan, M. and Cruz, J. (1973). Additional Aspects of the Stackelberg Strategy in Non-Zero Sum Games. *Journal of Optimization Theory and Applications* 11:613–626.

Chapter 8

A DIFFERENTIAL GAME THEORETIC ANALYSIS OF INTERNATIONAL TRADE IN RENEWABLE RESOURCES

With Hamid Beladi

We use a Stackelberg differential game to model trade in renewable resources between a monopsonistic buyer and a monopolistic seller. The buyer uses unit and *ad valorem* tariffs to indirectly encourage conservation of the resource under study. First, we show that the efficacy of these tariffs in furthering conservation depends essentially on whether harvesting costs are stock dependent or independent. Second, we study the impacts that alternate biological growth functions and the dependence of welfare in the buying country on the resource stock have on the optimal tariffs. Third, we note that because the simultaneous use of both tariffs does not render one tariff extraneous, it makes sense for the buyer to use both tariffs concurrently. Finally, we show that when the buyer uses both tariffs simultaneously, she can force the monopolistic seller to behave competitively.

1. Introduction

Can trade policies be used to further the conservation of renewable resources? As noted by Barbier *et al.* (1994) and Burgess (1994), this question has assumed great significance in contemporary times. To see why trade policies might be relevant in the context of the conservation of renewable resources, note that resources such as fish, timber, insects, ivory, and rhino horns are all commonly traded between nations. However, as Clark (1973), Jablonski (1991), and

Pimm *et al.* (1995) have pointed out, there is now considerable apprehension about the deteriorating stock levels of many salient renewable resources. As such, the general purpose of this chapter is to analyze the effects of trade policy on the conservation of renewable resources.

We now briefly discuss the pertinent literature. Barbier and Schulz (1997) show that trade interventions may increase or decrease the equilibrium value of the species stock in a developing country that trades in this species with other nations. Schulz (1997) shows that the threat of trade sanctions will not necessarily result in lower harvesting and higher stocks of marine mammals. Brander and Taylor (1998) show that tariffs imposed by a resource importing nation may benefit the resource exporter. Emami and Johnston (2000) argue that the trade induced losses that arise from not managing a fishery can be mitigated by imposing import tariffs on the resource good. Maestad (2001) shows that trade limitations may decrease or increase timber logging.

The efficacy of trade policies in furthering the conservation of renewable resources has been discussed most extensively in the context of the ivory trade between African countries and East Asian and western nations. Barbier *et al.* (1990) argue against a ban on trade in ivory and suggest an alternate strategy. In the aftermath of the Convention on International Trade in Endangered Species (CITES) ban on ivory trade, Bulte and van Kooten (1999) caution against lifting the trade ban. Heltberg (2001) has used a numerical model to reason that the ivory trade ban is likely to reduce poaching. This brief summary brings us to the first of our two specific points. The extant literature has *not* analyzed the connections between renewable resource harvesting costs and trade policy and the possible *time inconsistency* of optimal trade policies.

Our second specific point concerns the *differences* between exhaustible and renewable resources and the study of time inconsistency in the literature on exhaustible as opposed to renewable resources. There are two key differences. First, exhaustible resources do not regenerate but renewable resources do in accordance with some (stock dependent) biological growth function. Second, in contrast with exhaustible resources, renewable resources are often an argument in the utility functions of citizens in resource importing nations. Now, the problem of time inconsistency arises when agents with market power make promises that they would subsequently like to break. This problem — see Karp and Newbery (1993) and Groot *et al.* (2003) — has now been fairly well studied in the exhaustible resources literature. As noted in Karp and Newbery (1993, pp. 882–883), a key insight of this literature is that when (i) the future affects the present, (ii) at least one economic agent has market power and is able to influence the future, and (iii) the agent with market power cannot credibly commit herself to future actions, the problem of time inconsistency is salient. These three features are also present in the models that we analyze in this chapter. In addition, Karp and Newbery (1993, p. 892) note that the problem of time inconsistency is caused by stock dependent costs. Therefore, a key question that we analyze is whether stock dependent costs alone account for the time inconsistency of optimal policies or whether other factors can also cause time inconsistency.

The rest of this chapter is organized as follows: Section 2 provides a description of the Stackelberg differential game model. Section 3 studies trade policy when the renewable resource harvesting cost function is stock independent. Section 4 does the same when the harvesting cost function is stock dependent. Finally, Section 5 concludes and offers suggestions for future research on the subject of this chapter.

2. The Stackelberg Differential Game

As discussed by Anderson and Fong (1997) and Datta and Mirman (1999), international trade in renewable resources typically occurs in imperfectly competitive markets. In addition, the work of Edington and Hayter (1997) and Ruseski (1999) tells us that in these imperfectly competitive markets, strategic considerations are frequently very important. These two observations explain why we use a game theoretic model. Further, because renewable resources regenerate, a reasonable game theoretic model of such resources must be dynamic in nature. This explains our use of a differential game model. Finally, Brown and Layton (2001) and the references cited in this paper tell us that market power between the buyers and the sellers of renewable resources is frequently asymmetric.[1] This is why we use a Stackelberg differential game to model the interaction between a buyer and a seller of a renewable resource across international borders.

Our model is adapted from Karp (1984) and Batabyal (1996). There is a single buyer (the monopsonist and the leader) of the resource and this buyer purchases the resource from a single seller (the monopolist and the follower). Our primary aim is to study the problem of time inconsistency in the context of trade in renewable resources. Our purpose is not to analyze a general equilibrium model of all trade in a particular resource. This is why we analyze a partial equilibrium, two-country model of trade in a renewable resource. Moreover, this framework has been used previously (see Anderson and Fong (1997) and Okuguchi (1998)).

We now provide three examples of real world trade in a renewable resource that can be studied usefully with our Stackelberg game theoretic framework. First, suppose the resource is the black rhino horn. Then, following Brown and Layton (2001), our model can be thought of as a description of trade between a single seller such as Zimbabwe (follower) and a single buyer such as South Korea (leader). Second, consider an alternate resource, namely, salmon. In this case, following Schulz (1997), we can use our framework to analyze trade between a single seller such as Norway (follower) and a single buyer such as the USA (leader). Finally, suppose the resource is catfish. Then, following Kinnucan (2003), our framework can be used to analyze trade between Vietnam (follower) and the USA (leader). Although this is not our primary intent, our single buyer model can also be thought of as a model of a buyer's cartel. For

---

1. For more on this point, see the publications of the wildlife trade monitoring network Traffic International such as Milliken *et al.* (1993). Also useful are Baskin (1991), Dangar (2003), and recent issues of the periodical *African Wildlife*.

the rhino horn, the cartel might represent nations that use the horn primarily for medicinal purposes (China, South Korea, Singapore) or it might represent nations that use the horn mainly in ceremonial daggers (Oman, Saudi Arabia, Yemen).

The stock of the renewable resource at time t is $x(t)$. The buyer's utility from consuming harvest $h(t)$ is given by the concave and differentiable utility function $u(h)$. The domestic market in the buyer's country is competitive so that $u'(h) = p(h)$, the price that consumers in the importing nation pay for this resource. The government of the importing nation has access to a unit tariff $n(t)$ and an *ad valorem* tariff $a(t)$ where $v = 1/(1 + a)$. When the importing government uses $n(t)$, the price received by the monopolistic seller (exporter) is $p(h) - n(t)$. Similarly, when this government uses $a(t)$, the price received by the exporter is $v(t)p(h)$.[2]

The buyer's payoff is the discounted stream of the difference between the utility of consuming $h(t)$ and the payment to the monopolistic seller, from time $t = 0$ to $t = T$. Therefore, if we denote the interest rate by r, then the buyer's payoff is

$$J_b = \int_0^T e^{-rt}[u(h) - \{p(h) - n\}h]dt, \qquad (1)$$

when she uses a unit tariff. Similarly, when she uses the *ad valorem* tariff, her payoff is

$$J_b = \int_0^T e^{-rt}[u(h) - vp(h)h]dt. \qquad (2)$$

The monopolistic seller maximizes profit and he gets no utility from consuming the resource under study. A major aim of ours is to demonstrate the dependence of optimal trade policy on the form of the cost of harvesting the resource x. As such, we analyze two cost functions. The first cost function, analyzed in Section 3, is stock *independent* and it depends only on the harvest. This thrice differentiable cost function is $c(h)$, $c'(h) \geq 0$ and $c''(h) \geq 0$. The

---

2. Tariffs are arguably the most common trade policy instrument and, as such, their impacts have been studied in a variety of settings in the international trade literature and in the international trade and environment literature. For examples of such studies the reader should consult Batra et al. (1998), Batra and Beladi (1999), and Batra (2001).

second cost function, analyzed in Section 4, depends on the stock *and* on the harvest. Denote this function by $c(x)h$, where $c'(x) \leq 0$ and $c''(x) \geq 0$. The stock independent cost function is more relevant when the amount harvested is small in relation to the total size of the resource stock. In contrast, the stock dependent cost function is more relevant when the total size of the stock is small to begin with and/or when the amount harvested is a sizable proportion of the stock size. When the buyer uses $n(t)$, the seller's payoff is

$$J_s = \int_0^T e^{-rt}[p(h)h - nh - C(\cdot)]dt \qquad (3)$$

and when this buyer uses the *ad valorem* tariff, the seller's payoff is

$$J_s = \int_0^T e^{-rt}[vp(h)h - C(\cdot)]dt, \qquad (4)$$

where $C(\cdot)$ in Equations (3) and (4) is $c(h)$ and $c(x)h$, respectively.

The buyer controls the tariffs $n(t)$ and $v(t)$, and the seller controls the harvest $h(t)$. The buyer announces a tariff trajectory at the beginning of the game and this trajectory is exogenous to the seller. The buyer and the seller are constrained by the equation describing the dynamics of the resource stock. That equation is

$$\frac{dx(t)}{dt} = \dot{x} = f(x) - h(t), \qquad (5)$$

where $0 < k < r < \infty$ and $x(0) = x_0 > 0$ is given. Except in Section 3.3, for reasons to be explained in that section, we suppose that $f(x) = kx$. In Section 3.3, we analyze the logistic growth function where $f(x) = ax(1 - x/K)$, a is the intrinsic growth rate of the resource, and K is the carrying capacity.

3. The Stock Independent Harvest Cost Function and Open Loop Tariffs

We now derive the optimal unit and *ad valorem* tariffs for our single buyer. Although open loop tariffs are generally time inconsistent (Karp and Newbery, 1993; Batabyal, 1996), with two caveats that are discussed in Section 3.3, the open loop tariffs of this section *are* time consistent. If these tariffs were time inconsistent, then at some time $t > 0$ the buyer would want, if she could, to deviate from the tariff trajectory she announced at the beginning of the game and announce an alternate tariff trajectory. The monopolistic seller is forward looking. Therefore, he would anticipate the buyer's desire to change the tariff trajectory she announced at the beginning of the game and hence this tariff would fail to accomplish its intended goals. We now derive, in turn, the open loop unit and the *ad valorem* tariffs.

3.1. *The Open Loop Unit Tariff*

We follow Simaan and Cruz (1973), Karp (1984), and Batabyal (1996) to solve the buyer's problem. The buyer treats the seller's first-order necessary condition (FONC) as an ordinary constraint and his costate variable as a state variable. These two conditions and apposite boundary conditions convert the underlying differential game into a control problem for the buyer. When the seller takes $n(t)$ as given, the first-order necessary conditions to his problem are

$$p(h) + hp'(h) - c'(h) - \lambda - n = 0 \qquad (6)$$

and

$$\dot{\lambda} = (r - k)\lambda, \qquad (7)$$

where $\lambda(t)$ is the costate variable for (5). This variable gives us the seller's marginal utility of one more unit of the resource stock at time t. Equation (7) describes a jump state constraint (Karp and Newbery, 1993). That is, the initial value of λ, $\lambda(0)$, is free and the value of this jump state variable at any arbitrary point in time is determined by present and/or future events. Specifically, (7) is not a fixed initial state constraint for the buyer. Solving for n from (6) and substituting in (1), we get

$$J_b = \int_0^T e^{-rt}[u(h) - \{c'(h) + \lambda - hp'(h)\}h]dt. \qquad (8)$$

Equation (8) gives us the buyer's payoff. In particular, λh in (8) can be interpreted as the total instantaneous rent paid by the buyer for the resource x.

In order to keep the buyer's problem a one state variable problem, we eliminate λ from (8) by using (7). Solving (7), it is clear that $\lambda(t) = 0$. Substituting this value of 8 into (8) we get

$$J_b = \int_0^T e^{-rt}[u(h) - \{c'(h) - hp'(h)\}h]dt. \tag{9}$$

We have now converted the buyer's problem from one of maximizing (1) over n subject to (5) to one of maximizing (9) over h subject to (5). The FONCs to this problem are

$$p(h) + h^2 p''(h) + 2hp'(h) - hc''(h) - c'(h) - \eta = 0, \tag{10}$$

and

$$\dot{\eta} = (r - k)\eta, \tag{11}$$

where $\eta(t)$ is the costate variable for (5). Inspecting (10), we see that the solution to the buyer's problem does *not* depend on the initial stock of the resource x_0. This is why the optimal solution *is* time consistent. Put differently, because $\lambda(t) = 0$, the total instantaneous rent paid by the buyer for the resource x does not influence her maximization problem. Therefore, the question of altering the total instantaneous rent paid to the seller over time does not arise.

3.2. *The Open Loop Ad Valorem Tariff*

We now derive the solution for the open loop *ad valorem* tariff when the buyer's objective is (2). Specifically, we first maximize (4) subject to (5). The FONCs to the seller's problem are

$$v\{p(h) + hp'(h)\} - c'(h) - \lambda = 0 \tag{12}$$

and (7). Now solve for v from (12), substitute into (2), and then simplify. We get

$$J_b = \int_0^T e^{-rt}[u(h) - \{c'(h) + \lambda\}\varphi(h)]dt, \tag{13}$$

where $\varphi(h) = h/[1 + \{1/\theta(h)\}]$ and $\theta(h)$ is the price elasticity of demand. The buyer's problem now is to maximize (13) over h, subject to (5) and (7). The FONCs to this problem are

$$p(h) - \{c'(h) + \lambda\}\varphi'(h) - c''(h)\varphi(h) - \eta_1 = 0, \quad (14)$$

$$\dot{\eta}_1 = (r - k)\eta_1, \quad (15)$$

and

$$\dot{\eta}_2 = \varphi(h) + k\eta_2, \quad (16)$$

where η_1 and η_2 are the costate variables for (5) and (7). Inspecting Equations (14) through (16) we see that this solution does not depend on x_0. Hence, as in Section 3.1, this solution too is time consistent.

3.3. *Discussion*

Recently, Batabyal and Beladi (2006) have shown that when a single buyer of a renewable resource faces competitive sellers, irrespective of whether she uses a unit or an *ad valorem* tariff, her payoff is unchanged. Comparing Equations (13) and (8) we see that this result does not hold when a monopsonistic buyer trades with a monopolistic seller. Moreover, Batabyal and Beladi (2006) also tell us that when a single buyer faces competitive sellers, the optimal open loop unit and *ad valorem* tariffs are equivalent. Because the payoff functions (13) and (8) are not identical, in this chapter, the two tariffs are *not* equivalent.

Since the optimal unit and *ad valorem* tariffs are not equivalent, we would like to know what happens when our buyer uses both tariffs together. When the buyer uses both tariffs simultaneously, the seller maximizes

$$J_s = \int_0^T e^{-rt}[vp(h)h - nh - c(h)]dt, \quad (17)$$

over h, subject to (5). The FONCs to this problem include $n = v\{p(h) + hp'(h)\} - c'(h) - \lambda$. Substituting this last condition into the buyer's payoff function and then using $\lambda(t) = 0$ gives

$$J_b = \int_0^T e^{-rt}[u(h) + vh^2 p'(h) - hc'(h)]dt. \quad (18)$$

The buyer's current value Hamiltonian is linear and decreasing in v. Because $v \geq 0$, it is optimal to set $v = 0$ and this tells us that $a(t) = \infty$ and $n(t) = -c'(h)$. In other words, when our single buyer uses both tariffs simultaneously, it is optimal for her to levy an infinite *ad valorem* tariff and to offer a unit subsidy to the seller. Now substituting $v = 0$ in Equation (18) and then comparing the resulting equation with Equation (11) in Batabyal and Beladi (2006), we see that when our buyer uses both tariffs together, she is able to force the monopolistic seller to behave competitively. This means that the harvest rate of the resource is the same whether a competitive seller faces an optimal tariff of either kind or a monopolistic seller faces optimal unit and *ad valorem* tariffs. This last result arises because the simultaneous use of unit and *ad valorem* tariffs allows our buyer to shift and rotate the inverse demand function. As a result, she is able to confront the monopolistic seller with an infinitely elastic non-stationary function. This "rent appropriation" result has been obtained previously by Maskin and Newbery (1990) but in the context of exhaustible resources. To the best of our knowledge, our analysis is the first to show that this result carries over to the case of renewable resources as well. Finally, note that unlike the result obtained for the competitive case by Batabyal and Beladi (2006), the concurrent use of both tariffs does *not* make one tariff superfluous.

The aim of the importing nation is to encourage the conservation of the resource. Because the tariffs studied here are time consistent, they will achieve their intended conservation aims, albeit obliquely. Having said this, we should note that if an importing nation's goal is to encourage conservation of the renewable resource in the exporting country, then tariffs are not the ideal policy tools. This is because tariffs target trade and the direct consequence of the tariff is to discourage domestic consumption in the importing nation. Tariffs do not do anything directly to promote conservation of the resource in the exporting nation. Therefore, from the standpoint of resource conservation, tariffs are blunt policy instruments.

In our analysis thus far, we have shown that when the harvest cost function is stock independent, the optimal tariffs are time consistent. Is this result general? To answer this question, we now analyze two distinct cases. In the first case, the biological growth function in Equation (5) is logistic, i.e., $f(x) = ax(1-x/K)$. In the second case, the buying nation derives utility from consumption of the harvest and from the resource itself and hence $u(h)$ in Equations (1) and (2) is replaced by $u(h, x)$. Now, to demonstrate the impact of the logistic growth function on the optimal policy, let us redo the Section 3.1 analysis for the unit tariff. The FONCs to the buyer's problem now are

$$u'(h) + h^2 p''(h) + 2hp'(h) - c'(h) - hc''(h) - \lambda - \beta_1 = 0, \quad (19)$$

$$\dot{\beta}_1 = (r - a)\beta_1 + \{2a(\beta_1 x - \beta_2)\}/K, \quad (20)$$

and

$$\dot{\beta}_2 = a\beta_2 + h, \quad (21)$$

where β_1 and β_2 are the costate variables for Equation (5) and the constraint $\dot{\lambda} = (r - a)\lambda + (2ax)/K$. Now, rewriting and then integrating Equation (5) with $f(x) = ax(1 - x/K)$ gives

$$x(t) = x_0 + \int_0^t (ax - ax^2/K)\,dm - \int_0^t h(m)\,dm. \quad (22)$$

From Equation (19), the optimal harvest depends on β_1 and from Equation (20), β_1 depends on x. Therefore, substituting x from Equation (22) into Equation (20) we see that the initial resource stock x_0 affects the optimal solution given by Equations (19) through (21) and hence this solution is time *inconsistent*.

We now study the case in which the buying nation's utility is $u(h, x)$ and not $u(h)$. Once again, redoing the Section 3.1 analysis, the FONCs to the buyer's problem are

$$\frac{\partial u}{\partial h} + h^2 p''(h) + 2hp'(h) - hc''(h) - c'(h) - \delta = 0 \quad (23)$$

and

$$\dot{\delta} = (r-k)\delta - \frac{\partial u}{\partial x}, \quad (24)$$

where δ is the costate variable for Equation (5). Now, integrating Equations (5) and (24) and then simplifying, it can be shown that

$$\delta(t) = \frac{e^{rt}\{\delta_0 + \int_0^t e^{-(r-k)m}\{\frac{\partial u}{\partial x}\}dm\}}{x(t)/\{x_0 + \int_0^t e^{-km}h(m)dm\}}. \quad (25)$$

From Equation (23), the optimal harvest depends on δ and from Equation (25), δ depends on x_0. Therefore, it is clear that the optimal solution given by Equations (23) through (24) is time *inconsistent*.

We have now answered the question we posed a short while ago. Specifically, the result that with a stock independent harvest cost function, the optimal tariffs are time consistent, is *not* general. Even when the harvest cost function is stock independent, when either the biological growth function is logistic or when the buyer's utility depends on consumption and on the resource stock, the optimal tariffs are time inconsistent. We now focus on the Stackelberg game between our single buyer and seller given that the harvest cost function is stock dependent.

4. The Stock Dependent Harvest Cost Function and Open Loop Tariffs

When the harvest cost function is stock dependent, the optimal unit and *ad valorem* tariffs are time inconsistent. This means that at some time $t > 0$ the buyer will want to depart from the tariff trajectory she announced at time $t = 0$ and announce a different tariff trajectory. The monopolistic seller in this chapter is forward looking. Therefore, he will anticipate the buyer's desire to alter the tariff trajectory she announced at the beginning of the game and hence this tariff will fail to attain its intended conservation aims. To see this critical feature of the optimal policies clearly, we now derive the optimal open loop unit and *ad valorem* tariffs.

4.1. The Open Loop Unit Tariff

The stock dependent harvest cost function is $c(x)h$. Suppose the monopolistic seller takes the buyer's unit tariff $n(t)$ as given. Then, the FONCs to his problem are

$$p(h) + hp'(h) - n - c(x) - \lambda = 0 \qquad (26)$$

and

$$\dot{\lambda} = (r - k)\lambda + c'(x)h, \qquad (27)$$

where $\lambda(t)$ is the costate variable for (5). This variable gives us the seller's marginal utility of one more unit of the resource stock at time t. Comparing (27) with (7) it is clear that when the harvest cost function is stock dependent, $\lambda(t) \neq 0$. Now solving for n from (26) and then substituting into (1), we get

$$J_b = \int_0^T e^{-rt}[u(h) + h^2 p'(h) - \{c(x) + \lambda\}h]dt. \qquad (28)$$

Equation (28) says that the buyer's payoff is the present discounted stream of the utility of consuming the harvest $h(t)$ and the term $h^2 p'(h)$ less the sum of the cost and the term λh.

In order to keep the buyer's problem a single state variable problem, we now use (27) to eliminate λ from (28). Integrating (27) we get

$$\lambda(t) = e^{(k-r)(T-t)}\lambda(T) - e^{-(k-r)t}\int_t^T e^{(k-r)m} c'(x)h\,dm. \qquad (29)$$

Substituting this value of $\lambda(t)$ from (29) into (28) we get

$$J_b = \int_0^T e^{-rt}[u(h) + h^2 p'(h) - c(x)h - \{e^{(k-r)(T-t)}\lambda(T) - e^{-(k-r)t}$$

$$\times \int_t^T e^{(k-r)m} c'(x)h\,dm\}h]dt. \qquad (30)$$

Equation (30) tells us that for any trajectory of $h(t)$, it is optimal to set $\lambda(T) = 0$. By doing this, our buyer eventually drives the seller's rent to zero. Now substitute $\lambda(T) = 0$ into Equation (30) and then reverse the order of integration of the last term in (30). This gives us

$$J_b = \int_0^T e^{-rt}[u(h) + h^2 p'(h) - c(x)h + c'(x)h\{e^{kt}x_0 - x\}]dt. \quad (31)$$

We now have a single state variable control problem for our buyer. Specifically, we have converted the buyer's problem from one of maximizing Equation (1) over n subject to Equation (5) to one of maximizing Equation (31) over h subject to Equation (5). Comparing Equations (31) and (9) we see that unlike the case studied in Section 3.1, when the harvest cost function is stock dependent, x_0 enters the buyer's payoff function. The FONCs to our single buyer's problem are

$$p(h) + 2hp'(h) + h^2 p''(h) - c(x) + c'(x)\{e^{kt}x_0 - x(t)\} - \eta = 0, \quad (32)$$

and

$$\dot{\eta} = (r - k)\eta + 2hc'(x) - c''(x)h\{e^{kt}x_0 - x(t)\}, \quad (33)$$

where $\eta(t)$ is the costate variable. Equation (32) tells us that when the harvest cost function is stock dependent, the solution to the buyer's problem does depend on x_0. This has the following implication: At some time $m > 0$, if the buyer were able to revise the tariff she initially announced, then x_0 in Equation (32) would have to be replaced with $x(m)$ and the solution for all $t > m$ would be different. Put differently, the buyer's optimal solution is time *inconsistent*. From Equations (32) and (33) we see that the optimal solution is time consistent if and only if the harvest cost function is unrelated to the stock of the resource. Indeed, this is precisely what we demonstrated in Section 3.

A comment on the salience of Equation (27) is in order. When the seller exerts market power, he is subject to the sequence of static constraints implied by the buyer's optimizing behavior. These constraints enter the seller's problem as parameters and this allows him to solve a standard control problem. When the buyer exerts market power, she is constrained by the seller's dynamic optimizing behavior. As such, the buyer solves a non-standard control problem. As noted in Karp (1984) and in Karp and Newbery (1993), because the seller's problem is dynamic and the buyer exerts market power, a "rational expectations" constraint, given by Equation (27), is introduced into the buyer's problem. In Section 3.1, we called this "rational expectations" constraint a jump state constraint.

4.2. The Open Loop Ad Valorem Tariff

When the seller takes the buyer's *ad valorem* tariff as given, the FONCs to his control problem are

$$v\{p(h) + hp'(h)\} - c(x) - \lambda = 0, \tag{34}$$

and Equation (27). Solving for v from Equation (34), substituting into (2), and then simplifying, we get

$$J_b = \int_0^T e^{-rt}[u(h) - \{c(x) + \lambda\}\varphi(h)]dt, \tag{35}$$

where the functions $\varphi(h)$ and $\theta(h)$ are as indicated in Section 3.2. Comparing Equations (35) and (28) it is clear that these two payoff functions are not identical. Hence, unlike the competitive sellers case analyzed in Batabyal and Beladi (2006), the unit and the *ad valorem* tariffs are now *not* equivalent.

We now use the Section 3.2 method to analyze our buyer's problem. This buyer's problem is to maximize Equation (35) over h subject to Equations (5) and (27). The FONCs to this problem are

$$u'(h) - \varphi'(h)\{c(x) + \lambda\} - \eta_1 + \eta_2 c'(x) = 0, \tag{36}$$

$$\dot{\eta}_1 - (r - k)\eta_1 = \varphi(h)c'(x) - \eta_2 hc''(x), \tag{37}$$

and

$$\dot{\eta}_2 = \varphi(h) + k\eta_2, \tag{38}$$

where η_1 and η_2 are the costate variables for Equations (5) and (27). The difficulty in demonstrating the time inconsistency of the optimal solution (Equations (36) through (38)) arises from the presence of the generally nonlinear $\varphi(h)$ function in Equation (38). However, when the inverse demand function is isoelastic $\{p(h) = h^{-b}, b \in (0, 1)\}$, $\varphi(h)$ is linear and we can show the time inconsistency of the above optimal solution in a straightforward manner.

When $p(h) = h^{-b}, b \in (0, 1), \varphi(h) = h/(1 - b)$. Substituting this in Equation (38) and then solving the resulting differential equation gives us $(1-b)e^{-kt}\eta_2(t) = \int_0^t e^{-km}h(m)dm$. Consistent with Simaan and Cruz (1973), Karp and Newbery (1993), and Batabyal (1996), this solution uses the result that it is optimal to set $\eta_2(0) = 0$. Also, solving Equation (5), we get $x_0 - e^{-kt}x(t) = \int_0^t e^{-km}h(m)dm$. Equating these last two solutions we conclude that

$$\eta_2(t) = \frac{e^{kt}x_0 - x(t)}{1 - b}. \tag{39}$$

Now, inspecting Equation (36) we see that the optimal harvest of the resource is affected by η_2, the marginal value to the buyer of an increase in the seller's rent. In turn, from Equation (39), η_2 depends on x_0. Hence, the optimal harvest is itself a function of x_0, and following the logic of the argument of Section 4.1, this optimal solution, too, is time inconsistent. We note once again that the optimal solution (Equations (36) through (38)) is time consistent if and only if the harvest cost function is unrelated to the stock of the resource.

4.3. Discussion

Batabyal and Beladi (2006) have shown that when a monopsonistic buyer of a renewable resource faces competitive sellers, irrespective of whether she uses a unit or an *ad valorem* tariff, her payoff is unchanged. This result does *not* hold when the exporter is a monopolist, the importer is a monopsonist, and the harvest cost function is stock independent. Now, comparing Equations (35) and (28) we see two things. First, the Batabyal and Beladi (2006) "policy invariance" result also does *not* hold when the exporter is a monopolist, the importer is a monopsonist, and the harvest cost function is stock dependent. Second, because the payoff functions given by Equations (35) and (28) are not identical, the two tariffs themselves are now *not* equivalent.

We now show that when our buyer uses both tariffs concurrently, she can force the monopolistic seller to behave competitively. Since the logic of the argument is very similar to that employed in Section 3.3, we shall be brief. The FONC to the seller's maximization problem is $n = v\{p(h) + hp'(h)\} - c'(x) - \lambda$. Substituting this into the buyer's payoff function, we get

$$J_b = \int_0^T e^{-rt}[u(h) + vh^2 p'(h) - hc(x) - \lambda h]dt. \quad (40)$$

Inspecting Equation (40) we conclude that the buyer's current value Hamiltonian is linear and decreasing in v. Since $v \geq 0$, $v = 0$ is optimal and this tells us that $a(t) = \infty$ and $n(t) = -\{c(x) + \lambda\}$. In other words, when our single buyer uses both tariffs together, it is optimal for her to set an infinite *ad valorem* tariff and to offer a unit subsidy to the seller. Substituting $v = 0$ in Equation (40) and comparing the resulting equation with Equation (21) in Batabyal and Beladi (2006) we see that when our buyer uses both tariffs jointly, she is able to force the monopolistic seller to behave competitively. In other words, the harvest rate of the resource is the same whether a competitive seller faces an optimal tariff of either kind or a monopolistic seller faces optimal unit and *ad valorem* tariffs. Finally, unlike the competitive case studied in Batabyal and Beladi (2006), the synchronized use of both tariffs does *not* make one tariff extraneous.

Thus far we have shown that efforts to further resource conservation by means of trade policies are problematic in more ways than one. We now discuss this important point in greater detail. Karp and Newbery (1993) and Batabyal (1998) tell us that given a choice between time consistent and inconsistent policies, an economic agent will generally choose time inconsistent policies because inconsistent policies lead to a higher payoff. This means that even when the unit and *ad valorem* tariffs are time inconsistent, the buyer will prefer to use these inconsistent policies rather than follow a consistent course of action. Nevertheless, inconsistent policies are not believable and hence such policies will fail to attain their intended conservation objectives.

In contrast, when the harvest cost function is stock independent, the optimal tariffs are time consistent and hence plausible from the standpoint of the seller. Therefore, in this case, the buyer's tariffs will, albeit indirectly, attain their conservation aims. However, this result is not general. As we showed in Section 3.3, even when the harvest cost function is stock independent, when either the biological growth function is logistic or when utility in the buying nation depends on consumption and on the resource stock, the optimal tariffs are time inconsistent. Ideally, tariffs should not be used to promote the conservation of renewable resources. This is because tariffs reduce the domestic consumption of the traded resource in the importing nation and hence get at the conservation issue indirectly. However, if the seller is unwilling or unable to take measures in his own nation to further resource conservation, then tariffs are one imperfect instrument with which the seller can be encouraged to take the relevant conservation measures.

Although tariffs are an imperfect way of promoting conservation, they may not function as desired when they are needed. This is because the believability of the optimal tariffs depends on the form of the harvest cost function and this form is *not* controllable by the buyer. The extant literature in natural resource and environmental economics has *not* recognized this essential point. Now, the stock independent cost function is more suitable for relatively less threatened resources such as the catfish in Vietnam and the stock dependent cost function is more relevant for significantly more threatened resources such as the black rhino in Zimbabwe. Keeping in mind the Section 2 examples, our analysis suggests that the use of tariffs makes more sense in the context of the USA/Vietnam catfish trade and less sense in the context of the South Korea/Zimbabwe rhino horn trade. This suggestion appears to be consistent with available empirical evidence. As noted by Kinnucan (2003), tariffs have been used by the United States against catfish imports from Vietnam. Further, Brown and Layton (2001) tell us that prohibitive trade restrictions in the context of the rhino horn trade have generally failed to have any impact on the stock of the black rhino. One would think that for threatened resources, where the proper domestic conservation measures have not been taken, imperfect supra-national measures such as trade policies ought to be useful. However, this chapter's analysis tells us that trade policies are likely to be implausible and hence ineffectual for threatened renewable resource trade where the harvest cost function is typically stock dependent.

5. Conclusions

We conducted a Stackelberg game theoretic analysis of trade in renewable resources between a monopolistic seller and a monopsonistic buyer. Unit and *ad valorem* tariffs are used by the buyer to obliquely further the conservation of the resource under study. Unlike the finding contained in Batabyal and Beladi (2006), the optimal unit and *ad valorem* tariffs are *not* equivalent. Therefore, when the buyer uses both tariffs concurrently, she is able to dispense with the market power of the seller. The efficaciousness of trade policies in furthering resource conservation is, subject to the caveats noted in Sections 3.3 and 4.3, contingent upon the form of the harvest cost function. Our analysis shows that because of believability problems, tariffs are likely to be ineffectual in furthering the conservation of *threatened* renewable resources.

This chapter's analysis can be extended in a number of different directions. We suggest two possible extensions. First, given the time inconsistency of the optimal solution when the harvest cost function is stock dependent, it would be useful to compare and contrast the properties of time inconsistent and consistent tariffs in a differential game model of trade in renewable resources. Second, it would be interesting to study the trade game between a monopolistic seller and a fringe of competitive buyers and one dominant buyer. An examination of these aspects of the problem will allow richer analyses of the nexuses between international trade, trade policies, and the conservation of renewable resources.

References

Anderson, J.L. and Fong, Q.S.W. (1997). Aquaculture and International Trade. *Aquaculture Economics and Management* 1:29–44.

Barbier, E.B., Burgess, J.C., Swanson, T.M. and Pearce, D.W. (1990). *Elephants, Economics, and Ivory*. London, UK: Earthscan.

Barbier, E.B., Burgess, J.C., Bishop, J.T. and Aylward, B.A. (1994). *The Economics of the Tropical Timber Trade*. London, UK: Earthscan.

Barbier, E.B. and Schulz, C. (1997). Wildlife, Biodiversity, and Trade. *Environment and Development Economics* 2:145–172.

Baskin, Y. (1991). Archeologist Lends a Technique to Rhino Protectors. *BioScience* 41:532.

Batabyal, A.A. (1996). Consistency and Optimality in a Dynamic Game of Pollution Control II: Monopoly. *Environmental and Resource Economics* 8:315–330.

Batabyal, A.A. (1998). Environmental Policy in Developing Countries: A Dynamic Analysis. *Review of Development Economics* 2:293–304.

Batabyal, A.A. and Beladi, H. (2006). A Stackelberg Game Model of Trade in Renewable Resources with Competitive Sellers. *Review of International Economics* 14:136–147.

Batra, R. (2001). Are Tariffs Inflationary? *Review of International Economics* 9:373–382.

Batra, R. and Beladi, H. (1999). Trade Policies and Equilibrium Unemployment. *Manchester School* 67:545–556.

Batra, R., Beladi, H. and Frasca, R. (1998). Environmental Pollution and World Trade. *Ecological Economics* 27:171–182.

Brander, J.A. and Taylor, M.S. (1998). Open Access Renewable Resources: Trade and Trade Policy in a Two-Country Model. *Journal of International Economics* 44:181–209.

Brown, G. and Layton, D.F. (2001). A Market Solution for Preserving Biodiversity: The Black Rhino. *In* J.F. Shogren and J. Tschirhart (eds.), *Protecting Endangered Species in the United States*. Cambridge, UK: Cambridge University Press.

Bulte, E.H. and van Kooten, G.C. (1999). Economic Efficiency, Resource Conservation, and the Ivory Trade Ban. *Ecological Economics* 28:171–181.

Burgess, J. (1994). The Environmental Effects of Trade in Endangered Species. *In* OECD, *The Environmental Effects of Trade*. Paris: OECD.

Clark, C.W. (1973). The Economics of Overexploitation. *Science* 181: 630–634.

Dangar, A. (2003). Undercover in Yemen: The Rhino Trade. *Country Life* 197, Part 2:64–67.

Datta, M. and Mirman, L.J. (1999). Externalities, Market Power, and Resource Extraction. *Journal of Environmental Economics and Management* 37:233–255.

Edgington, D.W. and Hayter, R. (1997). International Trade, Production Chains, and Corporate Strategies: Japan's Timber Trade with British Columbia. *Regional Studies* 31:151–166.

Emami, A. and Johnston, R.S. (2000). Unilateral Resource Management in a Two-Country General Equilibrium Model of Trade in a Renewable Fishery Resource. *American Journal of Agricultural Economics* 82: 161–172.

Groot, F., Withagen, C. and de Zeeuw, A. (2003). Strong Time-Consistency in the Cartel-versus-Fringe Model. *Journal of Economic Dynamics and Control* 28:287–306.

Heltberg, R. (2001). Impact of the Ivory Trade Ban on Poaching Incentives: A Numerical Example. *Ecological Economics* 36:189–195.

Jablonski, D. (1991). Extinctions: A Paleontological Perspective. *Science* 253:754–757.

Karp, L.S. (1984). Optimality and Consistency in a Differential Game with Non-Renewable Resources. *Journal of Economic Dynamics and Control* 8:73–97.

Karp, L.S. and Newbery, D.M. (1993). Intertemporal Consistency Issues in Depletable Resources. *In* A.V. Kneese and J.L. Sweeny (eds.), *Handbook of Natural Resource and Energy Economics*, Volume 3. Amsterdam: Elsevier.

Kinnucan, H.W. (2003). Futility of Targeted Fish Tariffs and an Alternative. *Marine Resource Economics* 18:211–224.

Maestad, O. (2001). Timber Trade Restrictions and Tropical Deforestation: A Forest Mining Approach. *Resource and Energy Economics* 23:111–132.

Maskin, E. and Newbery, D.M. (1990). Disadvantageous Oil Tariffs and Dynamic Consistency. *American Economic Review* 80:143–156.

Milliken, T., Nowell, K. and Thomsen, J.B. (1993). *The Decline of the Black Rhino in Zimbabwe*. Cambridge, UK: Traffic International.

Okuguchi, K. (1998). Long-Run Fish Stock and Imperfectly Competitive International Commercial Fishing. *Keio Economic Studies* 35:9–17.

Pimm, S.L., Russell, G.J. and Gittleman, J.L. (1995). The Future of Biodiversity. *Science* 269:347–350.
Ruseski, G. (1999). Market Power, Management Regimes, and Strategic Conservation of Fisheries. *Marine Resource Economics* 14:111–127.
Schulz, C. (1997). Trade Sanctions and Effects on Long-Run Stocks of Marine Mammals. *Marine Resource Economics* 12:159–178.
Simaan, M. and Cruz, J.B. (1973). Additional Aspects of the Stackelberg Strategy in Non-Zero Sum Games. *Journal of Optimization Theory and Applications* 11:613–626.

Chapter 9

ENVIRONMENTAL POLICY IN DEVELOPING COUNTRIES: A DYNAMIC ANALYSIS

We construct a dynamic model of the environmental policy formulation process in a stylized developing country (DC). We analyze the employment and output effects of three pollution control policies. These policies embody different assumptions about the DC government's ability to commit to its announced course of action. We characterize the time path of the government's policy variable. We show that optimality calls for an activist policy, irrespective of the length of time to which the government can commit to its announced policy. This notwithstanding, the effects of this activist policy depend fundamentally on the government's period of commitment.

1. Introduction

In recent times, there has been considerable discussion on the general question of environmental policy in developing countries (DCs). There is general agreement (see Bhalla (1992), Mehmet (1995), and Renner (1992)) that a concerted attempt must be made by DC governments to design and implement policies which generate employment. However, in order to protect the environment, these same governments will also have to implement appropriate pollution control policies. The developed country experience with pollution control policies (see Christainsen and Tietenberg (1985)) tells us that these policies will often have a negative effect on employment. Consequently, DC governments may find it difficult to institute policies which ensure that the twin goals of employment creation and environmental protection are met. Given this state of affairs, concern has been expressed about a DC government's ability to realistically commit to environmental policy for any reasonable length of time. Indeed, some observers have noted that in the face of pressing employment creation needs, DC governments may not be serious about the question of environmental protection. Alternately put, although DC governments may initiate the process of establishing pollution control policies, their will to continue with such policies is likely to be limited.

Despite the importance of the issues discussed in the previous paragraph, the employment/environment question in DCs has received very little attention in the literature.[1] Therefore, the primary objective of this chapter is to construct and analyze an employment driven dynamic model of the environmental policy formulation process in a stylized DC. This chapter's secondary objective is to show how the DC government's optimal course of action is closely related to its ability to commit to its announced policy.

The rest of this chapter is organized as follows. Section 2 describes our theoretical framework in detail. Sections 3 through 5 analyze a dynamic model of the conduct of environmental policy by the government of a stylized DC, under three different assumptions about the ability of this government to commit to its announced policy. Section 6 offers concluding comments.

2. The Theoretical Framework

Our model is in the tradition of papers such as Mussa (1978), Pindyck (1982), and particularly Karp and Paul (1994), which study the implications of government/regulatory policies in a dynamic framework. We shall use the specific factors model to study a small, two-sector, trading DC. In order to stress the employment aspect of the underlying story, we shall assume that the DC economy is dualistic. In other words, the two DC sectors consist of a modern, high-wage, environmentally intensive sector in which production

---

1. See Lekakis (1991) and Mehmet (1995) for a more detailed corroboration of this claim.

causes pollution. The second sector is the traditional, low-wage, environmentally benign sector in which there is no pollution.

In order to earn higher wages, workers migrate from the traditional sector to the modern sector. This migration, which is unplanned from the perspective of the DC government, results in increased employment in the modern sector, increased production, and hence greater pollution.[2] Although workers, in their role as consumers, are adversely affected by pollution, they do not factor pollution into their migration decisions. As a result, the marginal migrant pays less than the marginal social cost of migration. In other words, in the absence of governmental policy, migration takes place too quickly and hence there is excessive pollution in the economy. In this situation, the first best policy is to tax pollution directly. However, in many DCs the government simply does not possess the wherewithal to tax pollution directly. As such, in this chapter we shall assume that the DC government operates in a second-best environment in which it controls pollution by means of a production tax.

Initially, the DC economy is in disequilibrium, owing to the fact that the government does nothing to correct distorted producer incentives and hence ensure environmental protection. A movement toward equilibrium requires a reduction in the production of the polluting good over time. Alternately put, a move toward equilibrium involves slowing the rate at which workers migrate from the traditional sector to the modern sector. We assume that workers have rational expectations, which is equivalent to assuming that they have perfect foresight in this deterministic model.

Each sector of the DC produces a single good using a fixed factor and a mobile factor called labor, with decreasing returns to scale. Superscripts on production variables will denote the sector and superscripts on consumption variables will denote the agent. Subscripts will denote partial derivatives. $L^i(t), i = 1, 2$ is the labor

---

[2]. In addition to having an adverse environmental impact, unplanned migration can be problematic in other ways as well. For more on this, see Swaminathan (1993).

employed by the ith sector at time t; time is continuous. Let $\bar{L}$ denote the DC's total labor endowment, i.e., $L^1(t) + L^2(t) = \bar{L}$. Good 2 is the polluting good. All our subsequent results are independent of whether good 2 is the export good or the import competing good. The government's policy variable is a production tax, $\tau(t)$, which is levied on the production of good 2. Following Dixit and Norman (1980) and Karp and Paul (1994), we shall use duality theory to model consumption and production decisions in the DC. The production function of the ith sector, $i = 1, 2$ is $f^i(L^i)$ and the corresponding revenue function is $R^i(p^i, L^i)$. As is well known, $R_1^i(\cdot)$ and $R_2^i(\cdot)$ denote the output supply of good i and the wage in sector i, respectively.[3] Let the world price of good 2 be $p = p^2/p^1$, where $p^1 = 1$. Further, let $L^2 = L$, and let $L^1 = \bar{L} - L$.

There is a continuum of identical workers in each sector of the DC economy and a single capitalist is the residual claimant. We shall assume that all agents have homothetic preferences. Then, following Dixit and Norman (1980, p. 326), the expenditure function of agent $j, j = 1, 2, 3$ can be written as $\mathscr{E}(1, p, u^j) = U^j \mathscr{E}(p)$, where $\mathscr{E}(p)$ is the unit expenditure function and U^j is agent j's real income. National income for the DC is $U \equiv (\bar{L} - L)U^1 + LU^2 + U^3$, where the superscript 3 refers to the capitalist.

Let $m(t)$ denote the private value of migration for any worker at time t, i.e., $m(t)$ denotes the discounted value of the wage differential between the high-wage polluting sector and the low-wage non-polluting sector. Mathematically,

$$m(t) = \int_t^\infty e^{-r(s-t)} \{R_2^2(\cdot) - R_2^1(\cdot)\} ds, \qquad (1)$$

where r is the discount rate. The integral equation in (1) can be converted into a differential equation. This equation is

$$\frac{dm}{dt} = \dot{m} = rm + R_2^1(\cdot) - R_2^2(\cdot). \qquad (2)$$

---

3. For more on the properties of these dual functions, see Dixit and Norman (1980, Chapter 2).

A worker will migrate to the modern sector if and only if the private value of migration, $m(t)$, is at least as high as the private cost of migration. However, because workers do not factor pollution into their migration decisions, the social cost of migration is not equal to the private cost of migration. Let the social cost of migration be quadratic, i.e., $C(\dot{L}) = \alpha(\dot{L})^2$, where $\alpha > 0$. Note that we are thinking of migration as the change in the labor stock. Now, because the average social cost, $\alpha\dot{L}$, is less than the marginal social cost, $2\alpha\dot{L}$, in the absence of governmental intervention, migration for high wage employment takes place too rapidly, thereby increasing environmental degradation. To capture the fact that the private cost of migration is lower than the social cost, suppose that workers base their migration decision on a fraction $\delta, \delta \in (0,1)$, of the marginal social cost $2\alpha\dot{L}$. Alternately put, the migrating workers do not internalize the environmental externality stemming in part from their decision to migrate. Now, equating the private value of migration with the private cost of migration, we get an equation for the dynamics of labor migration. This equation is

$$\dot{L} = \frac{m}{2\alpha\delta}. \qquad (3)$$

Since this DC economy is open and because we are not allowing for the possibility of international borrowing, in equilibrium, trade must be balanced. That is

$$D(U, L, m, \tau) \equiv U\mathscr{E}(p) + \frac{m^2}{4\alpha\delta^2} - R^1(1, \bar{L} - L) - R^2(p - \tau, L)$$
$$+ \tau R_1^2(p - \tau, L) = 0. \qquad (4)$$

The first term in this "balance of trade deficit" expression refers to consumption expenditures. Equation (3) tells us that $C(\dot{L}) = m^2/2\alpha\delta^2$. Hence, the second term of Equation (4) denotes the social cost of pollution. The third and the fourth terms give the value of production, and the fifth term denotes tax revenues. The tax revenues are distributed in lump-sum fashion.

We shall be particularly interested in studying the DC government's optimal dynamic environmental policy under three assumptions about its ability to commit to a particular course of action. In the first case, the government commits to a tax trajectory for an infinite period of time. The reader should interpret this infinite period of commitment as a case in which environmental protection is enshrined in the DC constitution.[4] When this is done, it does not matter which government is in office because the dictates of the constitution will have to be followed. In the second case, the DC government commits to a tax trajectory for a finite period of time. This finite period of commitment is more reasonable, and this finite period should be thought of as the length of time during which a particular government is in office. Unfortunately, in both these cases, the optimal tax policy is dynamically inconsistent. That is, the government announces a tax trajectory at time $t = 0$. However, at some time $\varepsilon > 0$, the government will want to deviate from the trajectory that it announced at $t = 0$. As a result, the government's announced policy at time $t = 0$ will not be credible. This means that forward looking workers will not believe that the government will actually carry through with its initially announced policy, and hence this policy will fail to accomplish its objectives.

Since the credibility of governmental policy has been an important issue in many developing countries, *a priori*, it would seem necessary to study the implications of the DC government following a dynamically consistent course of action.[5] This is the third case that we shall study. In this scenario, the government commits to a tax trajectory for an infinitesimal period of time. In the limiting case in which the period of commitment approaches zero, the government's tax policy is time consistent. This completes our discussion of the theoretical framework. We now turn to the DC government's problem when it can commit to its tax policy for an arbitrarily long period of time.

---

4. If the DC in question were India, this period would be 1976. This is because until 1976, environmental protection did not figure anywhere in the Indian constitution. See Batabyal (1993) for further details.

5. Recall the Section 1 discussion about the concern as to the lack of commitment in DC governmental policies. In this connection, also see Fanelli *et al.* (1992).

3. Environmental Policy with Perfect Commitment

In this case, the DC government is able to make a binding commitment and choose its tax trajectory over $[0, \infty]$ at $t = 0$. In the language of control theory, this is the government's open loop tax policy. The open loop pollution tax is a function of calender time only. Workers have perfect foresight and they are forward looking. As discussed earlier, because the economy is in disequilibrium at $t = 0$, the initial value of $L, L(0) = L_0$, does not equal the steady state value of labor in the polluting sector of the economy. It is important to note that the private value of migration at any time $t, m(t)$, is determined by the current and the future values of the tax. In other words, the constraint represented by Equation (2) is a jump state constraint.[6] This means that the initial value of m, $m(0)$, is endogenous to the problem. Note that this feature of the model makes the government's problem a non-standard control problem. In this setting, the government's objective is to solve

$$\max_{\{U,\tau\}} \int_0^\infty e^{-rt} U ds, \qquad (5)$$

subject to Equations (2), (3), and (4), with initial condition $L(0) = L_0$. The current value Hamiltonian for this problem is

$$H = U - \lambda \left[U\mathscr{E}(p) + \frac{m^2}{4\alpha\delta^2} - R^1(\cdot) - R^2(\cdot) + \tau R_1^2(\cdot) \right]$$
$$+ \sigma_1 \left\{ \frac{m}{2\alpha\delta} \right\} + \sigma_2 \{rm + R_2^1(\cdot) - R_2^2(\cdot)\}, \qquad (6)$$

where λ is the Lagrange multiplier on constraint (4), and σ_1 and σ_2 are the costate variables corresponding to constraints (3) and (2),

---

6. Many problems in economics are characterized by the existence of jump states. In monetary economics, the exchange rate is generally a jump state because it is affected by current interest rates and agents' expectations of the future money supply. For more on jump state constraints, see Batabyal (1996a; 1996b), Karp and Newbery (1993), and Karp and Paul (1994).

respectively. The first-order necessary conditions are

$$\lambda = 1/\mathscr{E}(p), \tag{7}$$

$$\lambda\{\tau R_{11}^2(\cdot) - 2R_1^2(\cdot)\} + \sigma_2 R_{21}^2(\cdot) = 0, \tag{8}$$

$$\dot{\sigma}_1 = r\sigma_1 + \sigma_2 h(\cdot) + \lambda\{d(\cdot) + \tau R_{12}^2(\cdot)\}, \tag{9}$$

and

$$\dot{\sigma}_2 = \frac{\lambda m}{2\alpha\delta^2} - \frac{\sigma_1}{2\alpha\delta}, \tag{10}$$

where $d(\cdot) \equiv R_2^1(\cdot) - R_2^2(\cdot)$, and $h(\cdot) = R_{22}^1(\cdot) + R_{22}^2(\cdot)$. That is, $-d(\cdot)$ denotes the current private value of migration, and $h(\cdot)$ denotes the sum of the slopes of the marginal product of labor in the two sectors. Note that $h(\cdot) = \partial\{-d(\cdot)\}\,\partial L < 0$.

Our main interest lies in characterizing the optimal pollution tax trajectory and the magnitude of the optimal pollution tax. To this end, denote steady state values by the superscript S. Then from Equation (3), it follows that $m^S = 0$. From Equation (2), $d^S(\cdot) = 0$. Equation (10) implies that $\sigma_1^S = 0$. From Equation (8), it follows that $\sigma_2^S = [-\lambda\{\tau R_{11}^2(\cdot) - 2R_1^2(\cdot)\}/R_{21}^2(\cdot)]^S$. From Equation (9), $\sigma_2^S = [-\lambda\tau R_{12}^2(\cdot)/h(\cdot)]^S$. Setting these last two expressions equal, we get $\tau^S = [2R_1^2(\cdot)h(\cdot)/\{R_{11}^2(\cdot)h(\cdot) - R_{12}^2(\cdot)R_{21}^2(\cdot)\}]^S$. From Equation (8), it follows that $\tau(t) = [\{2\lambda(t)R_1^2(\cdot) - \sigma_2(t)R_{21}^2(\cdot)\}/\{\lambda(t)R_{11}^2(\cdot)\}]$. Because $m(0)$ is free, as Simaan and Cruz (1973) have noted, the appropriate boundary condition for σ_2 is $\sigma_2(0) = 0$. In other words, the DC government chooses its tax trajectory in such a way so that the social shadow value of m is zero at the beginning of the program. Using $\sigma_2(0) = 0$, it follows that $\tau(0) = 2R_1^2(\cdot)/R_{11}^2(\cdot)$. We can now state the following proposition.

Proposition 1.

(i) *Suppose that the revenue function in sector 2 is separable in its arguments. Then the optimal program with perfect commitment involves setting $\tau(0) = \tau(t) = \tau^S > 0$.*

(ii) *Suppose that the supply function for good 2 is linear, upward sloping, and that the cross partial derivatives of the sector 2 revenue* function are positive. Then the optimal program with perfect commitment sets $\tau(0) > \tau^S > 0$.

(iiii) *Suppose that the sector 2 revenue function is arbitrary. Then optimality calls for setting $\tau(0) > \tau^S > 0$, as long as $[R_1^2(0)/R_{11}^2(0)] > [R_1^2/\{R_{11}^2 - (R_{12}^2 R_{21}^2)/h(\cdot)\}]^S$.*

Proof.

Case (i): Separability of the revenue function implies that $R_{12}^2(\cdot) = R_{21}^2(\cdot) = 0$. Substituting this into the expressions for $\tau(0)$, $\tau(t)$ and τ^S, the claimed result follows.

Case (ii): If the supply function of good 2 is linear and upward sloping, then $R_{11}^2(\cdot)$ is constant. Using this fact along with $R_{12}^2(\cdot) > 0$, $R_{21}^2(\cdot) > 0$ in the tax expressions above, the claimed result follows.

Case (iii): It is straightforward to check that if $[R_1^2(0)/R_{11}^2(0)] > [R_1^2/\{R_{11}^2 - (R_{12}^2 R_{21}^2)/h(\cdot)\}]^S$, then it is optimal to set $\tau(0) > \tau^S > 0$.

□

Proposition 1 describes the nature of the DC government's optimal dynamic policy under certain specific conditions. Under the sufficient case (i) condition, there are no price or labor interaction effects. Consequently, the government's optimal course of action is to set a pollution tax that is constant over the entire length of the program. On the other hand, under the sufficient case (ii) and the necessary and sufficient case (iii) conditions, the optimal policy involves beginning with a "bang." In these two cases, the DC government moves toward an equilibrium by starting with a large pollution tax. It then gradually lowers this tax to the steady state level. Note that the government's open loop tax policy calls for an activist course of action. In other words, it is typically not optimal to set a zero tax at any point in the program. The intuitive reason for this is as follows. In this open loop case, there are no welfare losses from being unable to commit, because the open loop tax policy incorporates perfect commitment. As such, the case for doing nothing, which potentially arises when the government cannot commit, is ruled out. Hence, the government corrects for the domestic distortion, and its tax policy is activist.

While Proposition 1 provides conditions for a constant and a declining tax trajectory, these are not the only possible trajectories. If the conditions described in Proposition 1 do not hold, it is possible for the pollution tax to exhibit more complex dynamic behavior. Specifically, it is possible for the tax to exhibit non-monotonic behavior. This tells us that the pursuit of open loop policies can lead to taxes which exhibit complicated dynamic behavior.

If the DC government's optimal tax policy, as described in Proposition 1, is believed by all agents in the economy, particularly by the migrating workers, then this policy will achieve its objectives. That is, the pollution tax will reduce output and employment in sector 2 and slow the rate of migration from the non-polluting sector 1 to the polluting sector 2. However, these objectives will not be met because the government will have an incentive to deviate from the policy that it announced at $t = 0$. To see this, note that for any initial value of L, $L(0) \neq L^S$, the optimal initial shadow value of $m(t)$, $\sigma_2(0)$, is zero. However, because $\delta < 1$, on the announced tax trajectory, $\sigma_2(0) \neq 0$. As a result, at any time $\varepsilon > 0$, the government will want to deviate from the tax trajectory it announced at $t = 0$, and announce a new trajectory. In other words, the government's open loop tax policy is dynamically inconsistent. This means that unless there is some mechanism by which the DC government can be bound to its initially announced tax trajectory, this government will fail to achieve its initially announced employment and environmental objectives.[7]

From a practical perspective, this case of perfect commitment is clearly implausible because no government can realistically be expected to commit to its policy for an infinite period of time. Consequently, we now analyze the case in which the DC government is able to commit to its announced policy at $t = 0$, for a finite period of time only. This is the limited commitment case.

---

7. The extent to which the government will fail to achieve its objectives depends on the nature and the direction of deviation from the initially announced tax policy.

4. Environmental Policy with Limited Commitment

Given that governments are in office for a finite period, the most reasonable period of commitment would seem to correspond to the length of time during which a particular government is in office. Hence, we now study the case of limited commitment in which the government commits to a policy for $T \in \mathbb{R}_{++}$ time periods.

When the period of commitment is finite, an analysis of the equilibrium trajectory of the pollution tax is made complex because the resulting equilibrium depends on the manner in which agents form their expectations. If migrating workers base their expectations of future taxes on the history of taxes, then multiple equilibria are possible. As such, there is a sort of "generic indeterminacy" to the outcome of the imposition of the pollution tax. To eliminate this indeterminacy, we shall restrict attention to smooth Markov perfect equilibria. That is, the agents' decision rules, at any time t, depend only on the current value of the state (i.e., the stock of labor) and not on the manner in which the current state was reached. A candidate for an equilibrium is said to be perfect if this candidate is an equilibrium for any possible subgame (i.e., for any possible level of the stock of labor). In particular, whether or not some agents have deviated from their equilibrium strategies in the past, the continuance of these strategies represents equilibrium behavior on the part of all the agents involved.[8] From a practical standpoint, the Markov assumption is useful because it makes the DC government's equilibrium tax policy insensitive to agents' mistakes.

With this restriction of Markov perfection, the equilibrium that emerges when the government makes a commitment for T time periods can be characterized. At time periods $0, T, 2T, 3T, \ldots$, successive governments choose their own tax trajectories. That is, at each iT, $i = 0, 1, \ldots$, the ith government completes its term in office, and a new government chooses its tax trajectory for the next T time periods. At the end of T time periods, each government

---

8. Markov perfect equilibria are sometimes known as strong Markov perfect equilibria. The word "strong" emphasizes the fact that current decisions are based *only* on the current state and *not* on actions undertaken in any previous time period.

bequeaths L_T, the current labor stock, to its successor government. This government then pursues its environmental policy for the next T time periods, and so on.

With this interpretation of the limited commitment case, let $V(L)$ be the value of the government's program when its period of commitment is T periods and when the initial level of labor in the polluting sector is L. Now, the government solves

$$V(L) \equiv \max_{\{\tau, U\}} \int_0^T e^{-rt} U dt + e^{-rT} V(L_T), \qquad (11)$$

subject to Equations (2), (3), and (4). Note that $V(L_T)$ is a bequest function. This function denotes the value of the labor stock bequeathed by an arbitrary government to its successor. Also note that problem (11) is the same as the problem described in Section 3, with the exception that the government's period of commitment is now T as opposed to infinity. This means that the boundary conditions at the horizon of the program will be different, although the first-order necessary conditions themselves remain as in Equations (7) through (10).

As before, the fact that $m(0)$ is free tells us that it is optimal to choose the tax trajectory so that $\sigma_2(0) = 0$. Using this last condition in Equation (8), $\tau(0) = [2R_1^2(\cdot)/R_{11}^2(\cdot)]$. To determine $\tau(T)$, let $M(L)$ be the equilibrium current value of m, which is determined by the solution to Equation (11).[9] In this case, we can write $V(L) \equiv \bar{V}\{L, M(L)\}$ for some function $\bar{V}\{\cdot\}$. At the beginning of a specific time period iT, $i = 0, 1, \ldots$, it is clear that $\sigma_2(iT) = 0$. Further, the assumed smoothness of the value function gives $\sigma_2 = \partial \bar{V}\{\cdot\}/\partial M$ (Karp and Paul, 1994, p. 1388). That is, the social shadow value of M is equal to the marginal value of M in the bequest. Finally, the transversality condition for σ_2 is $\sigma_2(T) = \partial \bar{V}\{\cdot\}/\partial M = 0$. Using this condition in Equation (8), $\tau(T) = [2R_1^2(\cdot)/R_{11}^2(\cdot)]$. We can now state the following proposition.

---

9. The properties of this endogenous function of the state have been discussed elsewhere and hence we shall omit an elaborate discussion. For further details, see Karp and Newbery (1993) or Karp and Paul (1994).

Environmental Policy in Developing Countries: A Dynamic Analysis

Proposition 2.

(i) *Suppose that the sector 2 revenue function is separable in its arguments. Then the optimal program with limited commitment has $0 < \tau(0) = \tau(T) = \tau(t), t \in (0, T)$.*

(ii) *Suppose that the sector 2 revenue function is arbitrary. Then the optimal program with limited commitment has $0 < \tau(0) = \tau(T) \neq \tau(t), t \in (0, T)$.*

Proof.

Case (i): Separability of the revenue function implies that $R_{12}^2(\cdot) = R_{21}^2(\cdot) = 0$. Substituting this into the expressions for $\tau(0), \tau(t)$, and $\tau(T)$, the claimed result follows.

Case (ii): The expressions for $\tau(0)$ and $\tau(T)$ tell us that these two taxes are positive and that they are not functions of the mixed partial derivatives of the sector 2 revenue function. In contrast to this, $\tau(t)$ is a function of the mixed partial derivatives of the revenue function. Hence, we have $0 < \tau(0) = \tau(T) \neq \tau(t)$, $t \in (0, T)$. □

Under the conditions specified in Proposition 2, an optimal program once again calls for an activist pollution control policy. The DC government sets a positive pollution tax even though it can commit to its announced policy for only a finite period. A comparison of the first case in Propositions 1 and 2 tells us that when the sector 2 revenue function is separable in its arguments, whether commitment is perfect or limited has no bearing on the government's optimal course of action. Put differently, the separability of the revenue function is a sufficient condition for the nature of commitment not to matter. In contrast, a comparison of case (iii) of Proposition 1 and case (ii) of Proposition 2 tells us that the time paths of the pollution tax in these two programs are quite different when the revenue function is arbitrary. In particular, while the perfect commitment case calls for starting with a high tax and then lowering this tax to its steady state value, the limited commitment case calls for equalizing the tax at the beginning and at the end of the program.

Note the important role played by the endogenous function of the state, $M(L)$. This function performs the role of an "expectations" function. When the DC government solves its optimization problem taking this expectations function as given, the optimal tax trajectory results in an initial value of m, $m(0)$, which satisfies $m(0) = M\{L(0)\}$. That is, in equilibrium, every agent's point expectations are fulfilled. Further, this same optimal tax trajectory results in a terminal value of m such that $\partial \bar{V}(\cdot)/\partial M = \sigma_2(T) = 0$. As indicated earlier, at the horizon of the program, the shadow value of the state M equals the marginal value of M in the bequest function, and these two values equal zero.

While this limited commitment scenario is quite plausible, this equilibrium too is dynamically inconsistent. To see why, think of the Markov perfect case just studied as one in which an infinite sequence of governments conduct environmental policy during a time period of length T. Further, denote the tenure of each government in this infinite sequence by $\{iT\}_{i=0}^{\infty}$. As long as $T > 0$, each government behaves consistently at each i but *not* within a period of length T. Put differently, the government begins its tenure in office with the best of intentions, but some time later it will want to renege on the policy it announced at the beginning of its tenure. As a result, forward looking agents will not believe that the government will actually carry through with its initially announced policy. In turn, this means that the government will not succeed in accomplishing its policy objectives. Pollution and employment in sector 2 will not be reduced, and the rate of migration from the traditional sector to the modern sector will not be slowed.

So far, we have shown that the dynamic inconsistency of the government's optimal tax policy is responsible for the non-attainment of the DC government's employment and environmental goals. This suggests a need to make the government's policy dynamically consistent. We now proceed to do this by studying the case in which the DC government commits to a specific tax policy for an infinitesimal period. In this setting, we shall be particularly interested in the limiting Markov perfect equilibrium in which the government's period of commitment shrinks to zero.[10]

5. Environmental Policy with Infinitesimal Commitment

Intuitively, one would expect the Markov perfect equilibrium to depend on the government's period of commitment. That is, one would expect the government's equilibrium tax to be a function of two opposing forces. The first force, the presence of pollution, would appear to necessitate an activist policy designed to correct for this domestic distortion. The second force, the government's inability to commit to its tax trajectory, would appear to favor the *status quo*. Given this scenario, the relevant policy question is this: Are there circumstances in which the welfare loss from being unable to commit dominates the welfare gain from reducing pollution, so that it is optimal to do nothing?[11]

In order to study the limiting case, we shall follow Karp and Newbery (1993) and Karp and Paul (1994) and begin with a discrete stage formulation of the DC government's problem. Let the government's period of commitment, and the length of each stage, be ε. Further, suppose that all agents act at the beginning of each time period of length ε. The state constraints facing the government at any time t can be written as

$$L_t = \left\{ \frac{m_t}{2\alpha\delta} \right\} \varepsilon + L_{t-\varepsilon} \qquad (12)$$

and

$$m_t = e^{-r\varepsilon} m_{t+\varepsilon} - d_t(\cdot)\varepsilon, \qquad (13)$$

---

10. For an alternate approach to the construction of dynamically consistent policies, see Batabyal (1996a; 1996b).

11. In this context, doing nothing refers to the case in which the DC government sets a zero tax.

where $d_t(\cdot) \equiv R_2^1(\cdot) - R_2^2(\cdot)$. In Equation (12), $(m_t/2\alpha\delta)\varepsilon$ represents the number of migrants in a period of length ε. Similarly, in Equation (13), $-d_t(\cdot)\varepsilon$ denotes the value of the flow of the wage differential in a time period of length ε.[12] At time t, with period of commitment ε, the government's dynamic programming problem is

$$\max_{\{U,\tau\}}[U - \lambda\{D(U,L,m,\tau)\}]\varepsilon + e^{-r\varepsilon}V(L_t), \qquad (14)$$

subject to Equations (12) and (13). Note that the function $D(\cdot)$ represents the "balance of trade deficit" constraint embodied in Equation (4), that $m_{t+\varepsilon} = M(L_t)$, and that the government takes the function $M(\cdot)$ as given. After some algebra, the first-order necessary condition with respect to τ can be written as

$$\lambda\left[\tau R_{11}^2(\cdot) - 2R_1^2(\cdot) - \left\{\frac{\partial D(\cdot)}{\partial L_t}\frac{dL_t}{d\tau}\right\} - \left\{\frac{\partial D(\cdot)}{\partial m_t}\frac{dm_t}{d\tau}\right\}\right]\varepsilon$$
$$+ e^{-r\varepsilon}\left\{\frac{dV}{dL_t}\frac{dL_t}{d\tau}\right\} = 0. \qquad (15)$$

In order to simplify Equation (15) further, it will be necessary to differentiate Equations (12) and (13) totally. This gives

$$\frac{dL_t}{d\tau} = \frac{\varepsilon}{2\alpha\delta}\frac{dm_t}{d\tau} \qquad (16)$$

and

$$\left\{-h_t(\cdot)\varepsilon - e^{-r\varepsilon}\frac{dM(\cdot)}{dL_t}\right\}\frac{dL_t}{d\tau} + \frac{dm_t}{d\tau} = -\left\{\frac{\partial d_t(\cdot)}{\partial \tau}\right\}\varepsilon. \qquad (17)$$

Now substitute for $dL_t/d\tau$ from Equation (16) into Equation (17), and then simplify the resulting equation. This gives $dm_t/d\tau \sim O(\varepsilon)$. Similarly, substituting for $dm_t/d\tau$ from Equation (17) into Equation (16) and then simplifying the resulting equation yields $dL_t/d\tau \sim o(\varepsilon)$. Now divide both sides of Equation (15) by ε, use the preceding two results regarding $dm_t/d\tau$ and $dL_t/d\tau$, and then

---

12. Equations (12) and (13) represent constraints (3) and (2) in discrete form.

let $\varepsilon \to 0$. The limiting first-order necessary condition becomes

$$\lambda[\tau R_{11}^2(\cdot) - 2R_1^2(\cdot)] = 0. \tag{18}$$

This leads to the following proposition.

Proposition 3. *In an optimal program in which the government's period of commitment $\varepsilon \to 0$, the limiting Markov perfect equilibrium tax is positive and equals $2R_1^2(\cdot)/R_{11}^2(\cdot)$.*

Proof. Since $\lambda > 0$, Equation (18) can be simplified to yield $\tau = [2R_1^2(\cdot)/R_{11}^2(\cdot)] > 0$. □

Proposition 3 provides us with an answer to the policy question posed at the beginning of this section. This proposition tells us that even when the government displays no commitment to its tax policy, the welfare loss from being unable to commit is never as great as the welfare gain from reducing pollution. Consequently, the optimal pollution tax is positive. Put differently, the passive aspect (do nothing) of governmental policy is dominated by the activist aspect (control pollution). This explains why the limiting pollution tax is positive.

While Proposition 3 and the discussion preceding it provide a rigorous characterization of the limiting pollution tax, the same characterization can be obtained intuitively. To see this, recall the discussion of the government's optimal policy immediately preceding the statement of Proposition 2. According to this discussion, $\tau(iT) = 2R_1^2(\cdot)/R_{11}^2(\cdot)$, $i = 0, 1, 2, 3, \ldots$. Now note that by choosing T sufficiently small, the equilibrium tax trajectory can be kept arbitrarily close to its initial value which equals $2R_1^2(\cdot)/R_{11}^2(\cdot)$. Not surprisingly, this is also the value of the limiting pollution tax.

This limiting case involves continuous revision of the pollution tax by the DC government. When the government revises its policy continually, the resulting policy is dynamically consistent. In other words, the government's tax policy is credible. Hence, this policy will be successful in reducing pollution in the modern sector and in slowing the rate at which workers migrate from the traditional sector to the modern sector.

As Karp and Newbery (1993) have noted, the payoff to an agent is monotonic in his period of commitment. In the context of this chapter, this means that reducing the government's period of commitment can never make the government better off. With this observation and the previous discussion of policy efficacy in mind, it is possible to rank the three policies in terms of (i) the government's preference and (ii) the policy's ability to achieve its goals. From the DC government's perspective, the most desirable policy is the open loop policy; this policy allows the government to make a commitment for an infinite period. The second best policy is the Markov perfect tax policy with a finite period of commitment. The least desirable policy is the limiting Markov perfect tax policy. As contrasted to this ranking, the ranking in terms of goal attainment is exactly the opposite. The limiting Markov perfect tax policy is credible; as such, this policy will be able to reduce pollution. The other two policy instruments are not credible; hence they will fail to achieve the government's environmental goals. Of these two non-credible policies, the finite commitment Markov perfect tax is more plausible. This discussion highlights the DC government's basic dilemma. The policy which results in the highest payoff to the government is the one that is least desirable from the standpoint of goal attainment and social welfare.

6. Conclusions

First, our analysis in this chapter provides some answers to hitherto little studied questions about the employment/environment interface in DCs. In particular, the analysis tells us that as long as the private cost of migration is less than the social cost of migration ($\delta < 1$), successful environmental policy involves continuous revision of the relevant policy instrument (tax).

Second, the analysis shows that doing nothing (i.e., setting a zero tax) is typically *not* an optimal course of action. We demonstrated that, although the nature of the underlying equilibrium depends on the government's ability to commit to its announced policy, the welfare loss from the inability to commit does not dominate the welfare gain from reducing pollution. As such, an optimal course of action generally requires that the pollution tax be positive.

Third, our analysis points to the unrealistic nature of dynamically inconsistent, particularly open loop, policies. Such policies cannot be believed by forward looking agents with rational expectations. Hence, such agents will successfully thwart the DC government's policy objectives. This stands in sharp contrast to the limiting Markov perfect tax policy which is dynamically consistent. In this case, the equilibrium is characterized by an endogenous function of the state and the government continuously revises its tax trajectory. Continuous revision implies credibility, and, in turn, this means that the government's environmental policy will achieve its intended objectives.

Fourth, there is a basic tradeoff between policy credibility and policy payoff. Credible policies yield a lower payoff than do noncredible policies. This observation provides a possible explanation as to why many DC governments are loath to use dynamically consistent policies which involve continuous policy revision.

The analysis contained in this chapter can be extended in a number of directions. One such extension would involve making the migration Equation (3) depend on m and the tax τ. This would permit an analysis of policy when the government's policy has a direct effect on migration decisions. Preliminary research along this line suggests that the DC government's optimal course of action can be quite sensitive to the manner in which the migration decision is modeled.

References

Batabyal, A.A. (1993). Should Large Developing Countries Pursue Environmental Policy Unilaterally? *Indian Economic Review* 28:191–202.

Batabyal, A.A. (1996a). Consistency and Optimality in a Dynamic Game of Pollution Control I: Competition. *Environmental and Resource Economics* 8:205–220.

Batabyal, A.A. (1996b). Consistency and Optimality in a Dynamic Game of Pollution Control II: Monopoly. *Environmental and Resource Economics* 8:315–330.

Bhalla, A.S. (1992). *Environment, Employment, and Development.* Geneva, Switzerland: International Labor Organization.

Christainsen, G.B. and Tietenberg, T.H. (1985). Distributional and Macroeconomic Aspects of Environmental Policy. *In* A.V. Kneese and J.L. Sweeney (eds.), *Handbook of Natural Resource and Energy Economics*, Vol. 1. Amsterdam, The Netherlands: Elsevier.

Dixit, A.K. and Norman, V. (1980). *Theory of International Trade.* Cambridge, UK: Cambridge University Press.

Fanelli, J.M., Frenkel, R. and Rozenwurcel, G. (1992). Trade Reform and Growth Resumption in Latin America. *In* J.-M. Fontaine (ed.), *Foreign Trade Reforms and Development Strategy.* London, UK: Routledge.

Kamien, M.I. and Scjwartz, N.L. (1991). *Dynamic Optimization*, 2nd edn. Amsterdam, The Netherlands: North-Holland.

Karp, L.S. and Newbery, D.M. (1991). Optimal Tariffs on Exhaustible Resources. *Journal of International Economics* 30:285–299.

Karp, L.S. and Newbery, D.M. (1993). Intertemporal Consistency Issues in Depletable Resources. *In* A.V. Kneese and J.L. Sweeney (eds.), *Handbook of Natural Resource and Energy Economics*, Vol. 3. Amsterdam, The Netherlands: Elsevier.

Karp, L.S. and Paul, T. (1994). Phasing in and Phasing Out Protectionism with Costly Adjustment of Labor. *Economic Journal* 104:1379–1392.

Lekakis, J.N. (1991). Employment Effects of Environmental Policies in Greece. *Environment and Planning A* 23:1627–1637.

Mehmet, O. (1995). Employment Creation and Green Development Strategy. *Ecological Economics* 15:11–19.

Mussa, M. (1978). Dynamic Adjustment in the Heckscher-Ohlin-Samuelson Model. *Journal of Political Economy* 86:775–792.

Pindyck, R.S. (1982). The Optimal Phasing of Phased Deregulation. *Journal of Economic Dynamics and Control* 4:281–294.

Renner, M. (1992). Jobs in a Sustainable Economy. *Worldwatch Paper No. 104.* Washington, DC.

Simaan, M. and Cruz, J.B. (1973). Additional Aspects of the Stackelberg Strategy in Non-Zero Sum Games. *Journal of Optimization Theory and Applications* 11:613–626.

Swaminathan, M.S. (1993). Ecotechnology and Rural Employment. *Environmental Conservation* 20:6–7.

Chapter 10

DYNAMIC ENVIRONMENTAL POLICY IN DEVELOPING COUNTRIES WITH A DUAL ECONOMY

With Dug Man Lee

We analyze a dynamic model of environmental policy in a stylized developing country (DC) with a dual economy. This DC's economy is distorted in part because the government subsidizes the exports of the non-polluting sector of the economy. We analyze the employment and output effects of three different pollution taxes. These taxes incorporate alternate assumptions about the DC government's ability to commit to its announced course of action. We describe the taxes, we examine the dependence of these taxes on the extant distortion, and we stipulate the conditions which call for an activist policy, irrespective of the length of time to which the government can commit to its announced policy. *Inter alia*, our analysis shows why some DC governments may not be serious about environmental protection.

1. Introduction

In contemporary times, issues at the interface of development and environmental economics have elicited a considerable amount of interest. On the one hand, researchers such as Bhalla (1992), Renner (1992), and Mehmet (1995) have argued that developing countries (DCs) need to take steps to design and follow through with policies that generate employment. On the other hand, the sizeable literature on sustainable development[1] has stressed the need for

---

1. For more on this literature, the reader should consult Brundtland (1987), Goldin and Winters (1995), Faucheux *et al.* (1996), and Atkinson *et al.* (1997).

implementing policies that will protect the environment for the present and future generations.

The experience of industrialized countries with environmental policies tells us that these policies can have a negative effect on employment (Christainsen and Tietenberg, 1985; Bonetti and FitzRoy, 1999). This and other similar findings have led many observers to contend that in the face of urgent employment creation[2] needs, DC governments are unlikely to be serious about environmental protection. In other words, although DC governments may begin the process of designing appropriate environmental policies, over time, their commitment to such policies is likely to wane.

The purpose of this chapter is to study this issue formally. We do this by analyzing a dynamic model of a DC with a dual economy. This model explicitly links the DC government's period of commitment to its announced employment/environmental policies. As Lekakis (1991) and Mehmet (1995) have noted, despite its significance, this issue has received scant attention in the economics literature.

Very recently, Batabyal (1998a) analyzed a dynamic model of environmental policy in DCs. In a two sector model, Batabyal (1998a) showed that the welfare gain from correcting for pollution is generally bigger than the welfare loss from being unable to commit to a particular environmental policy. As such, optimality calls for the DC government to conduct an activist environmental policy, regardless of the length of time to which this government can commit to its announced policy.

Historically, a number of DCs have protected their export sectors with subsidies to exporters. For instance, in the 1990s, Thailand provided subsidies on government-to-government sales of rice. Tajikistan subsidized the export of aluminum, and Venezuela granted bonuses to the exporters of certain kinds of agricultural products (Department of State, 1996). Given this state of affairs,

---

2. The reader may wish to think of employment creation policies as policies that increase the average wage in a DC.

we analyze the conduct of environmental policy in a DC in which the export — also the environmentally benign — sector is protected with an export subsidy. The specific question we address is the following. What are the properties of optimal environmental policy when a DC government controls pollution by taxing the production of the good manufactured by the polluting — also the import competing — sector, and when this government is unable to commit to the tax policy it announced at the beginning of its tenure in office? We show that an existing distortion in the export sector will have only a slight impact on a DC government's ability to conduct environmental policy effectively.

The rest of this chapter is organized as follows. Section 2 describes the theoretical framework in detail. Sections 3 through 5 examine a dynamic model of environmental policy in a DC, under three different assumptions about the ability of this DC government to commit to its initially announced policy. Section 6 concludes and offers suggestions for future research.

2. The Theoretical Framework

This chapter's model is in the tradition of Mussa (1978), Karp and Paul (1994), and Batabyal (1998a). These papers have all studied aspects of government policymaking in an intertemporal framework. We use a dynamic version of the specific factors model (also called the Ricardo–Viner model) to analyze a small DC.[3] This DC's economy is dualistic. One sector is the traditional, low wage sector in which there is no pollution. This traditional sector is also the export sector of the economy, and the DC government protects this sector by granting a subsidy to exporters. For political reasons, this subsidy cannot be repealed. Consequently, in the rest of this chapter,

---

3. Our focus on a small DC means that this DC's own policies do not affect world prices. As such, in the rest of this chapter, we shall not talk about the terms of trade effects of the DC government's environmental policies. As the reader will soon see, the principal focus of this chapter is on the time consistency/inconsistency of alternate pollution control policies. For more on the terms of trade effects of environmental policies, see Batabyal (1993; 1994a; 1994b).

we treat the subsidy as an exogenous parameter. The second sector is the modern, high wage sector in which production causes pollution. This pollution does not spill over into any other country. In other words, the pollution problem being analyzed in this paper is *not* transboundary in nature.[4] The modern sector is also the import competing sector of the DC.[5]

In order to earn higher wages, workers migrate from the traditional to the modern sector. This migration leads to increased employment in the modern sector. In turn, this increased employment leads to higher production, and hence to higher pollution. In their role as consumers, workers are detrimentally affected by pollution. Despite this, we suppose that they do not fully account for pollution in their migration decisions. This means that the marginal migrant pays less than the marginal social cost of pollution. In this setting, the first best policy is to tax pollution directly. However, in many DCs, governments do not have the ability to tax pollution directly. Therefore, we suppose that the DC government functions in a second best environment in which the best that it can do is to tax the production of the polluting good.[6]

Initially, the DC government does nothing to correct the distorted incentives faced by producers in the polluting sector. As such, the DC economy is in disequilibrium and the "balance of trade" account is unbalanced.[7] A movement toward equilibrium involves retarding the rate at which workers migrate from the traditional sector to the modern sector. Put differently, a move toward equilibrium requires a diminution in the production of the polluting good over

---

4. For more on the control of one kind of transboundary pollution, see Batabyal (1996; 1998b) and Xu and Batabyal (2001a; 2001b).

5. The reader may wish to think of the traditional sector as the agricultural sector, and the modern — possibly the infant industry — sector as the steel sector.

6. If the marginal social damage from pollution is such that the marginal damage from one unit of the polluting good is equal to the damage from one unit of pollution itself, then it does not matter whether the government taxes pollution directly or the production of the polluting good. However, in general, we would not expect this equality to hold. As such, our distinction between first and second best environments is germane.

7. We suppose that initially, the balance of trade account is in surplus. Batabyal (1998a) has analyzed the case in which initially, this account is in deficit.

time. We suppose that workers have rational expectations. In our deterministic model, this means that workers have perfect foresight.

Each sector of the DC produces a single good using a fixed factor and a mobile factor called labor, with decreasing returns to scale. Superscripts on production variables will denote the sector and superscripts on consumption variables will denote the agent. Subscripts will denote partial derivatives. Let $L^i(t), i = 1, 2$ denote the labor employed by the ith sector at time t; time is continuous. $\hat{L}$ denotes the DC's fixed labor endowment. This means that $\hat{L} = L^1(t) + L^2(t)$. Good 2 is the polluting — and the import competing — good. Let $\tau_s(t)$ denote the existing export subsidy in sector 1. The government's environmental policy instrument is a tax, $\tau_p(t)$, that is levied on the production of good 2.

Following Karp and Paul (1994) and Batabyal (1998a), we use duality theory to model consumption and production decisions in the DC. The production function in the ith sector, $i = 1, 2$, is $f^i(L^i)$. Let the price of the ith sector's good be p^i, and let $L^2 = L$, and $L^1 = \hat{L} - L$. Then we can write the revenue functions in the two sectors as $R^1(p^1 + \tau_s, \hat{L} - L)$ and $R^2(p^2 - \tau_p, L)$, respectively. Note that $R_1^i(\cdot,\cdot)$ and $R_2^i(\cdot,\cdot)$ denote the output supply of good i and the wage in sector i, respectively.[8]

A continuum of identical workers exist in each sector of the DC economy and a single capitalist is the residual claimant. All agents have homothetic preferences. Hence, the expenditure function of agent $j, j = 1, 2, 3$ can be written as $\bar{E}(p^1 + \tau_s, p^2, u^j) = U^j E(p^1 + \tau_s, p^2)$, where $E(\cdot,\cdot)$ is the unit expenditure function and U^j is the jth agent's real income. The DC's national income is $U \equiv (\hat{L} - L)U^1 + LU^2 + U^3$. The superscript j denotes the representative worker in sector j and $j = 3$ denotes the capitalist.

Let the private value of migration for any worker at time t be $m(t)$. In other words, $m(t)$ denotes the present value of the wage differential between the high wage polluting sector and the low

---

8. For more on the properties of these dual functions, the reader should consult Dixit and Norman (1980, Chapter 2).

wage non-polluting sector. Formally, we have

$$m(t) = \int_t^\infty e^{-r(s-t)}\{R_2^2(\cdot,\cdot) - R_2^1(\cdot,\cdot)\}ds, \qquad (1)$$

where r is the discount rate. This integral equation can be converted into a differential equation. That equation is

$$\frac{dm}{dt} = \dot{m} = rm + R_2^1(\cdot,\cdot) - R_2^2(\cdot,\cdot). \qquad (2)$$

We suppose that a worker will migrate to the modern sector if and only if the private value of migration, $m(t)$, is at least as high as the private cost of migration. Even so, because workers do not fully account for pollution in their migration decisions, the social cost of migration is not equal to the private cost of migration. For simplicity, we suppose that the social cost of migration is quadratic. Then $C(\dot{L}) = \alpha(\dot{L})^2$, with $\alpha > 0$. With this specification of the migration cost function, the average social cost of migration $\alpha\dot{L}$ is less than the marginal social cost $2\alpha\dot{L}$. This means that in the absence of governmental intervention, migration for high wage employment in the polluting sector takes place too rapidly, thereby increasing environmental degradation.

To account for the fact that the private cost of migration is less than the social cost of migration, suppose that the decision to migrate is based on a fraction $\delta, 0 < \delta < 1$, of the marginal social cost $2\alpha\dot{L}$. In other words, the migrating workers do not fully internalize the externality stemming in part from their decision to migrate. Now, if we equate the private value of migration with the private cost of migration, we get an equation for the dynamics of labor migration.[9] That equation is

$$\frac{dL}{dt} = \dot{L} = \frac{m}{2\alpha\delta}. \qquad (3)$$

---

9. If migration is affected by the private value of migration and by the pollution tax, then the expression for $\dot{L}$ in Equation (3) would have to be replaced by $\dot{L} = g(m, \tau_p)$ for some function $g(\cdot, \cdot)$. However, with this general functional form, it is not possible to derive closed form expressions for the optimal values of the pollution tax. This is the reason for working with the expression for $\dot{L}$ given in Equation (3).

Although our DC economy is open, we shall disallow the possibility of international borrowing. This means that in equilibrium, trade must be balanced. In other words,

$$D(U, L, m, \tau_s, \tau_p) = UE(\cdot, \cdot) + \frac{m^2}{4\alpha\delta^2} - R^1(\cdot, \cdot) - R^2(\cdot, \cdot)$$
$$+ \tau_s[R_1^1(\cdot, \cdot) - UE_1(\cdot, \cdot)] - \tau_p R_1^2(\cdot, \cdot) = 0 \quad (4)$$

must hold. The $D(\cdot, \cdot, \cdot, \cdot, \cdot)$ function is the "balance of trade" function. The first term on the RHS of Equation (4) denotes consumption expenditures. The second term on the RHS of Equation (4) — which equals $C(\dot{L})$ — denotes the social cost of pollution. The third and the fourth terms give the value of production. The fifth term denotes the value of the subsidy granted by the DC government to exporters of good 1. The sixth term denotes pollution tax revenue; as in Batabyal (1998a), this is assumed to be redistributed in lump-sum fashion.

Let us now study the DC government's optimal intertemporal environmental policy under three assumptions about its ability to commit to a particular course of action. In the first case, the government commits to its announced tax policy for an infinite period of time. The reader may want to interpret this infinite period of commitment as a case in which environmental protection is enshrined in the constitution. When this is done, it does not matter which government is in office because the constitution will have to be followed. In the second case, the DC government commits to its tax policy for a finite period of time. This finite period of commitment can be thought of as the length of time during which a particular government is in office. As we shall see, in both these cases, the government's optimal tax policy is time inconsistent. This means that at some time $\varepsilon > 0$, the government will want to depart from the tax policy it said it would follow at time $t = 0$. Because this government's pollution tax policy is time inconsistent, it is not credible. The unfortunate effect of this lack of credibility is that forward looking workers will not believe that the government will actually carry through with its initially proclaimed policy. As a result, this policy will fail to achieve its objectives.

Because the credibility of government policy has been a salient issue in many DCs, it is useful to study the consequences of following a time consistent course of action.[10] This is the third case that we study. In this case, the government commits to its tax policy for an infinitesimal period of time. In the limit, as this period of commitment tends to zero, the government's tax policy is time consistent. We now analyze the government's problem when it commits to its tax policy for an infinite period of time.

3. The Perfect Commitment Case

In this case, the DC government makes a binding commitment and selects its tax policy from time $t = 0$ to $t = \infty$, at $t = 0$. This is the government's open loop tax policy. The open loop pollution tax is a function of calendar time only. Workers have perfect foresight and they are forward looking. Because the economy is initially in disequilibrium, the initial value of L in the modern sector, $L(0) = L_0$, is not equal to its stationary state value. Further, because the decision to migrate is an investment decision, the private value of migration at time t, $m(t)$, is determined by the current and the future values of the pollution tax. The important consequence of this fact is that Equation (2) is actually a jump state constraint.[11] Mathematically, this means that the initial value of m, $m(0)$, is endogenous to the problem. In this setting, the DC government solves

$$\max_{\{U,\tau_p\}} \int_0^\infty e^{-rs} U ds, \qquad (5)$$

---

10. Recall the Section 1 discussion of the concern about the lack of commitment in DC government policies. For more on this, the reader should consult Fanelli *et al.* (1992).

11. For more on jump state constraints, the reader should consult Karp and Newbery (1993), Karp and Paul (1994), and Batabyal (1998a).

subject to constraints (2)–(4), with initial condition $L(0) = L_0$. The current value Hamiltonian for this problem is

$$H = U - \lambda \left[UE(\cdot,\cdot) + \frac{m^2}{4\alpha\delta^2} - R^1(\cdot,\cdot) - R^2(\cdot,\cdot) + \tau_s\{R_1^1(\cdot,\cdot) - UE_1(\cdot,\cdot)\} - \tau_p R_1^2(\cdot,\cdot) \right]$$

$$+ \sigma_1 \left\{ \frac{m}{2\alpha\delta} \right\} + \sigma_2 \{rm + R_2^1(\cdot,\cdot) - R_2^2(\cdot,\cdot)\}, \qquad (6)$$

where λ is the Lagrange multiplier on constraint (4), and σ_1 and σ_2 are the costate variables on L and m, respectively. The first-order necessary conditions are

$$\lambda = \frac{1}{E(\cdot,\cdot) - \tau_s E_1(\cdot,\cdot)}, \qquad (7)$$

$$\sigma_2 R_{21}^2(\cdot,\cdot) - \lambda \tau_p R_{11}^2(\cdot,\cdot) = 0, \qquad (8)$$

$$\dot{\sigma}_1 = r\sigma_1 + \lambda\{d(\cdot,\cdot) - \tau_s R_{12}^1(\cdot,\cdot) - \tau_p R_{12}^2(\cdot,\cdot)\} + \sigma_2 h(\cdot,\cdot), \qquad (9)$$

and

$$\dot{\sigma}_2 = \frac{\lambda m}{2\alpha\delta^2} - \frac{\sigma_1}{2\alpha\delta}, \qquad (10)$$

where $d(\cdot,\cdot) \equiv R_2^1(\cdot,\cdot) - R_2^2(\cdot,\cdot)$, and $h(\cdot,\cdot) \equiv R_{22}^1(\cdot,\cdot) + R_{22}^2(\cdot,\cdot)$. In other words, $-d(\cdot,\cdot)$ denotes the current private value of migration, and $h(\cdot,\cdot)$ denotes the sum of the marginal products of labor in the two sectors. Note that $h(\cdot,\cdot) = \partial\{-d(\cdot,\cdot)\}/\partial L < 0$.

The above first-order conditions can be used to shed light on the optimal pollution tax trajectory and to study the dependence of this tax on the export subsidy $\tau_s(t)$. To this end, let us denote stationary values by the superscript S. Equation (3) tells us that $m^S = 0$. Equation (2) reveals that $d^S(\cdot,\cdot) = 0$. Equation (10) gives $\sigma_1^S = 0$. From Equation (8) we get $\sigma_2^S = [\lambda \tau_p R_{11}^2(\cdot,\cdot)/R_{21}^2(\cdot,\cdot)]^S$ and Equation (9) yields $\sigma_2^S = [\lambda\{\tau_s R_{12}^1(\cdot,\cdot) + \tau_p R_{12}^2\}/h(\cdot,\cdot)]^S$. Setting these last two expressions equal gives $\tau_p^S = [\tau_s R_{12}^1(\cdot,\cdot) R_{21}^2(\cdot,\cdot)/\{R_{11}^2(\cdot,\cdot)h(\cdot,\cdot) - R_{12}^2(\cdot,\cdot)R_{21}^2(\cdot,\cdot)\}]^S$. From Equation (8), it follows that $\tau_p(t) = [\sigma_2(t)R_{21}^2(\cdot,\cdot)/\lambda(t)R_{11}^2(\cdot,\cdot)]$. As Simaan and Cruz (1973) have noted, because $m(0)$ is free, the proper boundary condition for σ_2 is $\sigma_2(0) = 0$. This means that the DC government selects its tax policy so that the initial social shadow value of m is zero. Using $\sigma_2(0) = 0$, we get $\tau_p(0) = 0$. Let us now write the three tax expressions compactly. We get

$$\tau_p(0) = 0, \quad \tau_p(t) = \frac{\sigma_2(t)R_{21}^2(\cdot,\cdot)}{\lambda(t)R_{11}^2(\cdot,\cdot)},$$

$$\tau_p^S = \left[\frac{\tau_s R_{12}^1(\cdot,\cdot)R_{21}^2(\cdot,\cdot)}{R_{11}^2(\cdot,\cdot)h(\cdot,\cdot) - R_{12}^2(\cdot,\cdot)R_{21}^2(\cdot,\cdot)}\right]^S. \quad (11)$$

3.1. *Discussion*

Recall that $h(\cdot,\cdot) < 0$. As discussed in Dixit and Norman (1980, pp. 35–43), the signs of the partial derivatives in Equation (11) are $R_{11}^i(\cdot,\cdot) > 0$, $R_{22}^i(\cdot,\cdot) < 0$, $R_{12}^i(\cdot,\cdot) > 0$, and $R_{21}^i(\cdot,\cdot) > 0$, for $i = 1, 2$. Further, because $\sigma_2(t)$ is the shadow value of $m(t)$, it will generally be positive. Finally, $\lambda(t)$, the multiplier on the balance of trade constraint, will also generally be positive. Now using this pattern of signs in Equation (11), we get $\tau_p(t) > 0$ and $\tau_p^S < 0$. This tells us that in an optimal program, the government begins with a zero tax. It then raises this tax over the length of the program, and then lowers the tax so that in the stationary state, the pollution tax is actually a subsidy.

To see why the initial pollution tax is zero, recall that in our second best environment, the purpose of this tax is to influence the rate of migration from the traditional to the modern sector. Now migration is an investment decision; as such, it involves balancing future benefits with current costs. A positive tax in the future discourages migration in the present period; call this the policy effect of the tax. In particular, the future tax moves the migration decision toward the socially optimal level. However, the future tax also leads to future distortions in production. Hence, from the standpoint of $t = 0$, a positive tax at $t > 0$ involves balancing a current benefit (the policy effect) with a future cost. This is exactly the opposite of the private migration decision which involves balancing future benefits with a current cost. A tax at $t = 0$ involves costs, but there are no previous migration decisions for this tax to influence. Put differently, a tax at $t = 0$ involves costs but it has no policy effect. This is why the optimal initial tax is zero.

Why is the optimal steady state pollution tax negative, i.e., a subsidy? To answer this question, recall that the purpose of the tax is to influence the rate of migration from the traditional to the modern sector, and observe from Equation (11) that the existing distortion (export subsidy) affects the optimal pollution tax at $t = \infty$ only, and at no other point in time. However, in the steady state, the private value of migration $m^S = 0$. This means that workers have no incentive to migrate from the traditional to the modern sector. Hence, from the standpoint of migration or pollution control, there is nothing to control. Put differently, the optimal pollution tax has no role to play. Despite this, the steady state level of labor in the two sectors is not optimal. This is because the positive export subsidy draws more than the optimal amount of labor into the traditional sector. As such, to encourage some migration from the traditional to the modern sector in the stationary state, and only in the stationary state, the government grants a subsidy to the producers of the polluting good. This subsidy ensures that the steady state level of labor in the two sectors is socially optimal.

In this open loop case that we have been discussing so far, there is no welfare loss to society from the government's inability to commit to its announced pollution tax policy. This is because, by definition, the open loop policy incorporates perfect commitment. This suggests that the following conjecture is true. Doing nothing, i.e., setting a zero pollution tax at all points in time, is a suboptimal course of action. Surprisingly, this conjecture is false. A sufficient condition under which this conjecture is false is when the revenue function in the polluting sector is separable in its arguments. In this case, $R^2_{12}(\cdot,\cdot) = R^2_{21}(\cdot,\cdot) = 0$, and it follows from Equation (11) that it is optimal to set $\tau_p(t) = 0$, $\forall t \in [0, \infty]$.

If the DC government's open loop tax policy is believed by the workers, then this policy will achieve its objectives. In particular, the pollution tax will reduce output and employment in sector 2 and slow the rate of migration from the traditional to the modern sector. However, the government's objectives will *not* be met because the government's open loop tax policy is time inconsistent. In other words, the government will have an incentive to depart from the policy it announced at $t = 0$. To see why, note that for any initial value of L, $L(0) \neq L^S$, the optimal initial shadow value of $m(t)$, $\sigma_2(t) = 0$. Nevertheless, because $\delta < 1$, on the announced tax trajectory $\sigma_2(t) \neq 0$. This means that at any time $\varepsilon > 0$, the government will want to depart from the tax trajectory it announced at $t = 0$ and announce a new trajectory. What this means is that in the absence of a device that can bind the government to its initially announced tax policy, this government will fail to achieve its initially announced employment and environmental objectives.

As noted by Batabyal (1998a), the case of perfect commitment is unpersuasive because no government can credibly commit to a policy for an infinite period of time. Consequently, we now examine the case in which the DC government commits to its initially announced policy for a finite period of time. This is the limited commitment case.

4. The Limited Commitment Case

Governments are generally in office for a limited amount of time. Consequently, the most sensible period of commitment equals the length of time during which a particular government is in office. So, let us now analyze the case in which the DC government commits to a policy for $T \in (0, \infty)$ time periods.

When the period of commitment is finite, an examination of the government's optimal program is complicated by the fact that the resulting equilibrium depends on the manner in which agents form their expectations. If workers base their expectations of future taxes on the history of taxes, then multiple equilibria can arise. To obviate this problem, we shall focus on smooth Markov perfect equilibria.

"Markov" means that the decision rules of the agents at time t depend only on the current value of the state (stock of labor) and not on the manner in which the current state was reached. A prospect for an equilibrium is perfect if this prospect is an equilibrium for any possible subgame (any possible level of the stock of labor). In particular, whether or not some agents have departed from their equilibrium strategies in the past, the continuation of these strategies represents equilibrium behavior for all the involved agents.

With this restriction of Markov perfection, we can now characterize the equilibrium that arises when the government commits to its tax policy for T periods. At points $0, T, 2T, \ldots$, successive governments select their tax policies. Put differently, at each iT, $i = 0, 1, 2, \ldots$, the ith government completes its tenure in office and a new government selects its tax policy for the next T periods. At the end of T periods, each government bequeaths the current stock of labor, L_T, to its successor government. This government then pursues its environmental policy for the next T periods, and so on.

Let $V(L)$ be the value of the government's program when its period of commitment is T and when the initial level of labor in the modern sector is L. The DC government solves

$$V(L) = \max_{\{U, \tau_p\}} \int_0^T e^{-rt} U dt + e^{-rT} V(L_T), \qquad (12)$$

subject to constraints (2)–(4). $V(L_T)$ is a bequest function. This function denotes the value of the stock of labor bequeathed by an arbitrary government to its successor. Note that except for the fact that the government's period of commitment is now T and not ∞, problem (12) is the same as problem (5). This means that the first-order necessary conditions to this problem will remain as in Equations (7)–(10). The boundary conditions, however, will change.

Because $m(0)$ is free, it is optimal to select the tax trajectory so that $\sigma_2(0) = 0$. Using this condition in Equation (8), we get $\tau_p(0) = 0$. As indicated in Equation (11), $\tau_p(t) = [\sigma_2(t) R_{21}^2 (\cdot, \cdot) / \lambda(t) R_{11}^2 (\cdot, \cdot)]$. To determine $\tau_p(T)$, let $M(L)$[12] denote the

---

12. For more details on this endogenous function of the state variable, the reader should consult Karp and Newbery (1993) and Karp and Paul (1994).

equilibrium current value of m that is determined by the solution to problem (12). Then, we can write $V(L) \equiv \hat{V}\{L, M(L)\}$, for some function $\hat{V}\{\cdot,\cdot\}$. At the beginning of a particular time period $iT, i = 0, 1, 2, \ldots$, we have $\sigma_2(iT) = 0$. Further, the assumed differentiability of the value function gives $\sigma_2 = \partial \hat{V}(\cdot,\cdot)/\partial M$ (Karp and Paul, 1994, p. 1388; Batabyal, 1998a, p. 15). This means that the social shadow value of $M(\cdot)$ is equal to the marginal value of $M(\cdot)$ in the bequest. Finally, the transversality condition for σ_2 is $\sigma_2(T) = \partial \hat{V}(\cdot,\cdot)/\partial M = 0$. Using this condition in Equation (8) gives $\tau_p(T) = 0$. Let us now write the three tax expressions compactly. We get

$$\tau_p(0) = 0, \quad \tau_p(t) = \frac{\sigma_2(t) R_{21}^2(\cdot,\cdot)}{\lambda(t) R_{11}^2(\cdot,\cdot)}, \quad \tau_p(T) = 0. \qquad (13)$$

4.1. Discussion

Comparing the three tax expressions in Equations (11) and (13), we see that a diminution in the length of the government's period of commitment has no impact on either $\tau_p(0)$ or $\tau_p(t)$. However, $\tau_p(T)$ differs from τ_p^S. This is because the transversality condition $\sigma_2(T) = 0$ — which does not apply in the perfect commitment case — can be used to simplify the expression for $\tau_p(T)$. An important implication of this is that the existing distortion (export subsidy) now has no effect on the DC government's optimal pollution tax policy. From the standpoint of policy efficacy, this means that distortions that are not in the polluting sector are far less likely to have a detrimental impact on the DC government's ability to conduct environmental policy effectively.[13]

---

13. Note that the results described in Equation (13) cannot be meaningfully compared with the results that would emerge from an analysis of environmental policy in a static Heckscher–Ohlin framework because the theoretical framework that generates Equation (13) is fundamentally different from a static Heckscher–Ohlin framework with an environmental externality. Specifically, the time consistency/inconsistency of policies is not an issue in the static Heckscher–Ohlin framework. In contrast, this issue is of central importance in the theoretical framework of this chapter.

As in Section 3, the DC government's optimal tax policy is not always activist. In particular, if the modern sector's revenue function is separable in its arguments, then $R_{21}^2(\cdot,\cdot) = 0$ and Equation (13) tells us that $\tau_p(t) = 0$, $\forall t \in [0, T]$. From the discussion in this section and in Section 3.1, the reader will note that when the modern sector's revenue function is separable in its arguments, whether commitment is perfect or limited has no bearing on the government's optimal course of action. Put differently, the separability of the modern sector's revenue function is a sufficient condition for the nature of commitment not to matter.

The endogenous function of the state, $M(L)$, performs the role of an "expectations" function. When the DC government solves its maximization problem taking $M(\cdot)$ as given, the ensuing optimal program leads to an initial value of m, $m(0)$, that satisfies $m(0) = M\{L(0)\}$. This means that in equilibrium, every agent's point expectations are fulfilled. Moreover, this same optimal program leads to a final value of m so that $\sigma_2(T) = \partial \hat{V}(\cdot,\cdot)/\partial M = 0$. What this means is that at the horizon of the program, the shadow value of the state M, equals the marginal value of M in the bequest function, and these two values equal zero.

Regrettably, even in this limited commitment case, the resulting Markov perfect equilibrium is time inconsistent. To see this, imagine an infinite sequence of DC governments conducting environmental policy for T periods. Denote the tenure of each government in this sequence by $\{iT\}_{i=0}^{\infty}$. Whenever $T > 0$, governments act consistently at each i, but *not* in a period of length T. One can think of governments that begin their tenure in office with the desire to act consistently, but then renege on the policy that they announced at the beginning of their tenure in office. Forward looking agents will not believe the policy announcements of such governments; consequently, the announced environmental policies of such governments will fail to achieve their employment and pollution control objectives.

In the preceding analysis, we have seen that the steady state dependence of the pollution tax on the export subsidy, and more importantly, the time inconsistency of the government's optimal tax policy will prevent the government from achieving its employment and environmental goals. This raises the following question. How can the government's optimal tax policy be made time consistent? To answer this question, we now examine the limiting Markov perfect equilibrium in which the government's period of commitment shrinks to zero.

5. The Infinitesimal Commitment Case

In general, we would expect the DC government's Markov perfect equilibrium tax to be a function of three elements. The first element — the presence of pollution — suggests that the government adopt an activist policy to correct for this external diseconomy. The second element — the government's inability to commit to its tax policy — would appear to favor the *status quo*. As we have seen, the third element — the export subsidy — is irrelevant except in the steady state. Given this state of affairs, the important policy question concerns the relative strengths of the first two elements. In other words, does the first element dominate the second so that it is optimal to set a positive pollution tax? Alternately, does the second element dominate the first so that it is optimal to set a zero pollution tax?

Note that by posing the policy question in these terms, we are comparing the benefit (reduced pollution) of a policy with its cost. In this chapter, we are abstracting away from implementation costs; as such, the cost under consideration is not the cost of carrying out a specific tax policy. Instead, this cost is the cost of announcing a particular course of action and then not following through with this announced course of action. We now answer the policy question posed at the end of the previous paragraph.

We follow Karp and Paul (1994) and Batabyal (1998a) and begin with a discrete stage formulation of the DC government's problem. Let the government's period of commitment and the length of each stage be ε. Further, suppose that all agents act at the beginning of each stage of length ε. Then at time t, the discrete versions of the two differential equation constraints — Equations (3) and (2) — facing the government are

$$L_t = \left\{\frac{m_t}{2\alpha\delta}\right\}\varepsilon + L_{t-\varepsilon}, \qquad (14)$$

and

$$m_t = e^{-r\varepsilon}m_{t+\varepsilon} - d_t(\cdot,\cdot)\varepsilon. \qquad (15)$$

In Equation (14), $\{m_t/2\alpha\delta\}\varepsilon$ denotes the number of migrants in a stage of length ε. Similarly, in Equation (15), $-d_t(\cdot,\cdot)\varepsilon$ represents the value of the flow of the wage differential in a stage of length ε.

At time t, with period of commitment ε, the government solves

$$V(L_{t-\varepsilon}) = \max_{\{U,\tau_p\}}[U - \lambda\{D(U,L,m,\tau_s\tau_p)\}]\varepsilon + e^{-r\varepsilon}V(L_t), \qquad (16)$$

subject to constraints (14) and (15). Here, $D(\cdot,\cdot,\cdot,\cdot,\cdot)$ denotes the balance of trade function described by Equation (4), $m_{t+\epsilon} = M(L_t)$, and the DC government takes the $M(\cdot)$ function as given. The first-order necessary condition to problem (16) w.r.t. τ_p is

$$\lambda\left[\tau_p R_{11}^2(\cdot,\cdot) + \frac{\partial D(\cdot,\cdot,\cdot,\cdot,\cdot)}{\partial m_t}\cdot\frac{dm_t}{d\tau_p} + \frac{\partial D(\cdot,\cdot,\cdot,\cdot,\cdot)}{\partial L_t}\cdot\frac{dL_t}{d\tau_p}\right]\varepsilon$$

$$- e^{-r\varepsilon}\frac{dV(\cdot)}{dL_t}\cdot\frac{dL_t}{d\tau_p} = 0. \qquad (17)$$

To simplify Equation (17), let us differentiate Equations (14) and (15) totally. This yields

$$\frac{dL_t}{d\tau_p} = \frac{\varepsilon}{2\alpha\delta}\cdot\frac{dm_t}{d\tau_p}, \qquad (18)$$

and

$$\left\{-h_t(\cdot,\cdot)\varepsilon - e^{-r\varepsilon}\frac{dM(\cdot)}{dL_t}\right\}\frac{dL_t}{d\tau_p} + \frac{dm_t}{d\tau_p} = -\left\{\frac{\partial d_t(\cdot,\cdot)}{\partial \tau_p}\right\}\varepsilon. \qquad (19)$$

Now substitute for $dL_t/d\tau_p$ from Equation (18) into Equation (19) and then simplify the resulting equation. This tells us that $dm_t/d\tau_p \sim O(\varepsilon)$. Similarly, substituting for $dm_t/d\tau_p$ from

Equation (19) into Equation (18) and then simplifying the ensuing equation gives $dL_t/d\tau_p \sim o(\varepsilon)$. Finally, divide both sides of Equation (17) by ε, use the results that $dm_t/d\tau_p \sim O(\varepsilon)$ and $dL_t/d\tau_p \sim o(\varepsilon)$, and then let $\varepsilon \to 0$. The limiting first-order necessary condition becomes

$$\lambda \tau_p R_{11}^2(\cdot,\cdot) = 0. \tag{20}$$

5.1. *Discussion*

As discussed in Section 3.1, in general, $\lambda > 0$ and $R_{11}^2(\cdot,\cdot) > 0$. Hence, in the general case, the limiting Markov perfect pollution tax is zero. This means that the answer to the question posed in the first paragraph of Section 5 is that in our model, the welfare loss from being unable to commit to environmental policy (the cost) swamps the welfare gain from reducing pollution (the benefit). As such, the limiting Markov perfect pollution tax is zero and it is optimal for the DC government to do nothing.

The limiting case that we analyzed in this section involves continuous revision of the pollution tax by the DC government. In addition to this, the limiting Markov perfect tax is not a function of the existing distortion (export subsidy). From this, we can draw the following conclusion. When the DC government revises the pollution tax continually, the government's environmental policy is time consistent and hence credible.

Karp and Newbery (1993) have remarked that the reward to an agent is monotonic in his period of commitment. As such, reducing the DC government's period of commitment can never make this government better off. This observation and the previous discussion of policy efficacy enables us to rank the three pollution taxes in term's of the government's preference, and the ability of the tax to accomplish its objectives. The government's reward is highest with the open loop pollution tax. Consequently, as indicated in Table 1, the government will prefer this policy the most. The next best policy is the Markov perfect pollution tax, with an intermediate level of commitment. The least desirable policy is the limiting Markov

Table 1. Properties and the rankings of the three pollution taxes.

Criteria	Open Loop Pollution Tax	Markov Perfect Pollution Tax	Limiting Markov Perfect Pollution Tax
Government's Period of Commitment	∞	$T \in (0, \infty)$	$\varepsilon > 0$
Time Consistent?	No	No	Yes
Ranking from Government's Perspective	1	2	3
Ranking from Goal Attainment Perspective	3	2	1

perfect pollution tax which involves continuous policy revision. In contrast with this ranking, Table 1 tells us that the ranking in terms of goal attainment is exactly the opposite. The limiting Markov perfect pollution tax is credible; as such, this policy will be able to slow migration to the polluting sector and reduce pollution. The open loop and the Markov perfect pollution taxes are time inconsistent. Hence, these policies will fail to achieve the government's employment and environmental objectives. In sum, from the standpoint of goal attainment, the most desirable policy is the limiting Markov perfect pollution tax.

This discussion starkly illustrates the DC government's dilemma. The policy which leads to the highest reward for the government is the one that is least likely to lead to the satisfaction of this government's policy goals. It is in this sense that the fear of observers, who have worried that in the face of urgent employment creation needs, DC governments are unlikely to be serious about environmental protection, is justified.

6. Conclusions

In this chapter, we examined the conduct of environmental policy by a DC government under three assumptions about this government's ability to commit to its announced policy. Four salient policy conclusions emerge. First, in general, the existing distortion has no effect on the optimal pollution taxes. This result tells us that the design of environmental policy in DCs should be informed by some knowledge of the sectors in which existing distortions lie. Distortions in the *non-polluting* sector are unlikely to have much of an impact on the government's ability to conduct environmental policy efficaciously.

Second, the inability to commit to an announced course of action is a serious roadblock to the conduct of successful environmental policy. Time inconsistent policies, particularly open loop policies, are impractical. Such policies will not be believed by forward looking agents with rational expectations. Hence, these agents will successfully thwart the DC government's policy objectives. This suggests that environmental protection should either be enshrined in the constitution, or it should be common knowledge that optimal pollution taxes will be revised frequently by the government.[14]

Third, in contrast with the finding contained in Batabyal (1998a), our analysis shows that in some situations, doing nothing, i.e., setting a zero pollution tax, *is* an optimal course of action. This is because in these situations, the welfare loss (the cost) from being unable to commit to a specific policy swamps the welfare gain (the benefit) from reducing pollution.

Fourth, there is a basic tradeoff between policy credibility and policy reward. Time consistent policies yield a lower reward than do time inconsistent policies. This observation provides a possible explanation as to why many DC governments are loath to use time consistent policies. Moreover, this observation also tells us that the

---

14. The primary focus of this chapter has been on the time consistency/inconsistency of alternate pollution control policies. As such, we have not addressed the question of optimal pollution abatement. However, note that continuous policy revision does not necessarily imply suboptimal investment in pollution abatement capital. Whether or not such investment is suboptimal will depend on the nature of agents' expectations. In particular, if these expectations are rational, i.e., if agents have perfect foresight (in a deterministic model), then, in principle, optimal investment in pollution abatement capital can go together with the pursuit of a time consistent course of action.

argument put forward by some observers that in the face of pressing employment creation needs, DC governments are unlikely to be serious about environmental protection, has some merit.

The analysis contained in this chapter can be extended in a number of directions. In what follows, we suggest two potential extensions. First, it would be helpful to determine what impact additional mobile factors of production have on a DC government's dynamic environmental policy. Second, following Batabyal (1997, 1999, 2000a, 2000b), one can study policy formulation in a broader setting in which the conduct of domestic environmental policy is warranted by a DC's participation in an international environmental agreement (IEA). Studies which incorporate these aspects of the problem into the analysis will provide richer accounts of the nexuses between existing distortions, time consistency, and environmental policy.

References

Atkinson, G., Dubourg, R., Hamilton, K., Munasinghe, M., Pearce, D. and Young, C. (1997). *Measuring Sustainable Development: Macroeconomics and the Environment*. Cheltenham, UK: Edward Elgar.

Batabyal, A.A. (1993). Should Large Developing Countries Pursue Environmental Policy Unilaterally? *Indian Economic Review* 28:191–202.

Batabyal, A.A. (1994a). An Open Economy Model of the Effects of Unilateral Environmental Policy by a Large Developing Country. *Ecological Economics* 10:221–232.

Batabyal, A.A. (1994b). On the Possibility of Attaining Environmental and Trade Objectives Simultaneously. *Environmental and Resource Economics* 4:545–553.

Batabyal, A.A. (1996). Game Models of Environmental Policy in an Open Economy. *Annals of Regional Science* 30:185–200.

Batabyal, A.A. (1997). Developing Countries and Environmental Protection: The Effects of Budget Balance and Pollution Ceiling Constraints. *Journal of Development Economics* 54:285–305.

Batabyal, A.A. (1998a). Environmental Policy in Developing Countries: A Dynamic Analysis. *Review of Development Economics* 2:293–304.

Batabyal, A.A. (1998b). Games Governments Play: An Analysis of National Environmental Policy in an Open Economy. *Annals of Regional Science* 32:237–251.

Batabyal, A.A. (1999). Developing Countries and Environmental Protection: Contract Design in Perfectly Correlated Environments. *Open Economies Review* 10:305–323.

Batabyal, A.A. (2000a). The Effects of Collusion and Limited Liability on the Design of International Environmental Agreements for Developing Countries. *In* A.A. Batabyal (ed.), *The Economics of International Environmental Agreements*. Aldershot, UK: Ashgate.

Batabyal, A.A. (ed.) (2000b). *The Economics of International Environmental Agreements*. Aldershot, UK: Ashgate.

Bhalla, A.S. (1992). *Environment, Employment, and Development*. Geneva, Switzerland: International Labor Office.

Bonetti, S. and FitzRoy, F. (1999). Environmental Tax Reform and Government Expenditure. *Environmental and Resource Economics* 13:289–308.

Brundtland, G.H. (1987). *The UN World Commission on Environment and Development: Our Common Future*. Oxford, UK: Oxford University Press.

Christainsen, G.B. and Tietenberg, T.H. (1985). Distributional and Macroeconomic Aspects of Environmental Policy. *In* A.V. Kneese and J.L. Sweeney (eds.), *Handbook of Natural Resource and Energy Economics*, Vol. 1. Amsterdam, The Netherlands: Elsevier.

Department of State (1996). *Country Reports on Economic Policy and Trade Practices*. Report submitted to the 104th Congress. Washington, DC: U.S. Government Printing Office.

Dixit, A.K. and Norman, V. (1980). *Theory of International Trade*. Cambridge, UK: Cambridge University Press.

Fanelli, J.M., Frenkel, R. and Rozenwurcel, G. (1992). Trade Reform and Growth Resumption in Latin America. *In* J.-M. Fontaine (ed.), *Foreign Trade Reforms and Development Strategy*. London, UK: Routledge.

Faucheux, S., Pearce, D. and Proops, J. (eds.) (1996). *Models of Sustainable Development*. Cheltenham, UK: Edward Elgar.

Goldin, I. and Winters, L.A. (eds.) (1995). *The Economics of Sustainable Development*. Cambridge, UK: Cambridge University Press.

Karp, L.S. and Newbery, D.M. (1993). Intertemporal Consistency Issues in Depletable Resources. *In* A.V. Kneese and J.L. Sweeney (eds.), *Handbook of Natural Resource and Energy Economics*, Vol. 3. Amsterdam, The Netherlands: Elsevier.

Karp, L. and Paul, T. (1994). Phasing in and Phasing Out Protectionism with Costly Adjustment of Labor. *Economic Journal* 104:1379–1392.

Lekakis, J.N. (1991). Employment Effects of Environmental Policies in Greece. *Environment and Planning A* 23:1627–1637.

Mehmet, O. (1995). Employment Creation and Green Development Strategy. *Ecological Economics* 15:11–19.

Mussa, M. (1978). Dynamic Adjustment in the Heckscher-Ohlin-Samuelson Model. *Journal of Political Economy* 86:775–791.

Renner, M. (1992). Jobs in a Sustainable Economy. *Worldwatch Paper No. 104.* Washington, DC: U.S. Government Printing Office.

Simaan, M. and Cruz, J.B. (1973). Additional Aspects of the Stackelberg Strategy in Non-Zero Sum Games. *Journal of Optimization Theory and Applications* 11:613–626.

Xu, Q. and Batabyal, A.A. (2001a). Price Competition, Pollution, and Environmental Policy in an Open Economy. *Annals of Regional Science* 35:59–79.

Xu, Q. and Batabyal, A.A. (2001b). A Bertrand Model of Trade and Environmental Policy in an Open Economy. *Keio Economic Studies* 38:53–70.

Chapter 11

A DYNAMIC ANALYSIS OF PROTECTION AND ENVIRONMENTAL POLICY IN A SMALL TRADING DEVELOPING COUNTRY

With Hamid Beladi

We analyze a dynamic model of protection and environmental policy in a small trading developing country (DC). The DC government protects the import competing (and the polluting) sector of the economy with a tariff. The employment and output effects of three different pollution taxes are analyzed. These taxes incorporate different assumptions about the DC government's ability to commit to its announced policy. First, we describe the taxes, we study the dependence of these taxes on the tariff, and we show that in general an activist environmental policy is called for, irrespective of the length of time to which the government can commit to its announced policy. Second, we identify a situation in which the conduct of environmental policy raises welfare unambiguously, and the situations in which it does not do so. Finally, we show that the time inconsistency of certain optimal programs can prevent the DC government from achieving its environmental and employment objectives.

1. Introduction

Four issues about environmental policy in developing countries (DCs) have increasingly come to dominate public debate in both the developing and the developed world. As Miller (1995) has noted, the first issue is the perception in many developed countries that DCs are not doing enough to protect their environmental resources.

The second issue — see Batabyal (1995) — concerns the potential effects of protection[1] and trade on environmental policy. Specifically, how should a trading DC conduct environmental policy when its protected import competing sector[2] is also the polluting sector?

The third issue concerns the need for creating employment opportunities in DCs. In this connection, Bhalla (1992), Renner (1992), and Mehmet (1995) have argued that DC governments must make a concerted attempt to design and implement policies that generate employment. The fourth issue concerns the apparent tradeoff between employment creation and environmental protection. Several researchers have noted that in addition to implementing employment creating policies, in order to protect the environment, DC governments will also have to implement environmental policies. The developed country experience with environmental policies tells us that these policies can have a negative effect on employment (Christainsen and Tietenberg, 1985; Bonetti and FitzRoy, 1999). This finding has led many to argue that in the face of pressing employment creation needs, DC governments are unlikely to be serious about environmental protection. Put differently, although DC governments may begin the process of instituting environmental policies, their commitment to such policies is likely to be limited.

To date, these issues have not been analyzed rigorously in the economics literature.[3] Consequently, in this chapter, we analyze these issues rigorously by formulating a dynamic model of environmental policy in a small trading DC. This model is distinct in the way in which it links a DC government's period of commitment to its announced environmental policies. Recently, Batabyal

---

1. The word "protection" refers to a situation in which one or more sectors of the economy are shielded from foreign competition by means of policies such as import tariffs. A tariff is a tax on an imported good.

2. The import competing sector of an economy competes with imported goods. For example, if a DC imports steel, then the domestic steel producing sector would be the import competing sector.

3. See Lekakis (1991), Mehmet (1995), and Batabyal (1998) for a more detailed corroboration of this claim.

(1998) has conducted a dynamic analysis of environmental policy in DCs. Using a simple model and specific assumptions about the form of the relevant revenue functions, Batabyal (1998) obtains results about the nature of optimal intertemporal environmental policy.

In this chapter, we substantially generalize this previous analysis. In particular, we make no assumptions about the form of the underlying revenue functions, and we study the conduct of environmental policy by a small trading DC in which a tariff protects the import competing and the polluting sector. The specific question that we address is the following: What are the properties of optimal dynamic environmental policy when a DC government controls pollution by taxing the production of the good manufactured by the protected sector, and when this government is not necessarily able to commit to the pollution tax policy it announced at the beginning of its tenure in office?

Four noteworthy outcomes follow from the more general analysis of this chapter. First, the model of this chapter is richer than the model in Batabyal (1998) and it includes the Batabyal (1998) model as a special case. Second, the model of this chapter is able to shed light on the following question: What effect does an existing distortion (the import tariff) have on the DC government's optimal dynamic environmental policy? This question cannot be answered using the model in Batabyal (1998). Third, this chapter's model answers the following question: What are the properties of optimal dynamic environmental policy in a second best environment?[4] This question too cannot be answered with the model in Batabyal (1998). Finally, the results that follow from an analysis of the more general model of this chapter are very different from the results in Batabyal (1998). In these ways, this chapter's model and analysis constitute a non-trivial improvement over Batabyal (1998) in particular and the extant literature in general.

---

4. By second best environment we mean a situation in which (i) the number of distortions exceeds the number of corrective policy instruments available to the DC government, and (ii) the DC government is unable to tax pollution directly. As we shall soon see, in this chapter there are two distortions (import tariff and pollution) and one policy instrument (pollution tax).

To the best of our knowledge, the issues of this chapter have not been analyzed previously in the operations research literature. Second, this chapter shows how "well-accepted operational research methods"[5] can be used to study interesting and salient problems at the interface of the environment and development. As such, in addition to being a novel contribution to the economics literature, we believe that our chapter is a new contribution to the operations research literature as well. Having said this, it should be noted that this chapter is intended to be a contribution primarily to the theoretical literature in operations research.

The rest of this chapter is organized as follows: Section 2 contains a detailed description of the theoretical framework. Sections 3 through 5 study a dynamic model of environmental policy by the government of a stylized DC, under three different assumptions about the ability of this government to commit to its initially announced policy. Section 6 offers concluding comments and suggests directions for future research.

2. The Theoretical Framework

Our model follows previous papers such as Mussa (1982), Karp and Paul (1994), and Batabyal (1998) that study government policies in a dynamic framework. We use a dynamic version of the Ricardo–Viner model[6] to study a small trading DC. To stress the employment aspect of the underlying story, we suppose that the DC economy is dualistic. This means that the two DC sectors consist of a modern, high wage, environmentally intensive sector in which production causes pollution. This polluting sector is also the import competing sector. The government uses a positive tariff to protect this sector. One possible interpretation of this sector is that it is the DC's

---

5. This language is taken from the comments of an anonymous referee.

6. This is a standard model in trade theory. In this model, there are two sectors with a factor of production specific to each of these two sectors and a mobile factor of production (typically labor) that can move between these two sectors. For more on this model, see Krugman and Obstfeld (2000, Chapter 3).

"infant industry."[7] The second sector is the traditional, low wage, environmentally benign sector that is free of pollution. This traditional sector — possibly the agricultural sector — is the DC's export sector. The political clout of the import competing sector is such that the government is unable to remove the tariff any time in the foreseeable future. As such, in what follows, we suppose that the tariff is exogenously given.[8]

To earn higher wages, workers migrate from the traditional sector to the modern sector. This migration results in increased employment in the modern sector, increased production, and hence greater pollution. In their role as consumers, workers are adversely affected by pollution. Nevertheless, in this chapter, they do not account for pollution in their migration decisions. This means that the marginal migrant pays less than the marginal social cost of migration. In this situation, ideally, one would want to tax pollution directly. However, in many DCs, the government simply does not possess the means to tax pollution directly. Consequently, we assume that the DC government functions in a second best environment in which it controls pollution with a production tax.

At first, the government does not correct the distorted incentives that producers face as a result of the presence of pollution. This is why the DC economy is initially in disequilibrium. This disequilibrium arises solely from the fact that, initially, there are distortions (pollution and the tariff) in the DC economy. There are no other factors that are responsible for this initial disequilibrium. Moreover, we suppose that this initial disequilibrium and all the subsequent equilibria that we analyze are stable. A move toward equilibrium requires that the production of the polluting good decline over time. Seen from a different perspective, a move toward equilibrium

---

7. An "infant industry" is a nascent indigenous industry. Initially, such industries frequently have high costs and hence they find it difficult to compete with other, more established, foreign industries. In turn, this difficulty is often the justification for the protection of "infant industries." For more on these issues, see Krugman and Obstfeld (2000, pp. 255–257).

8. Note that the purpose of this chapter is to study environmental policy when the polluting sector is protected; we are not interested in the dynamics of trade policy *per se*. This is the reason for treating the tariff as exogenous.

involves slowing the rate at which workers migrate from the traditional to the modern sector. Workers have rational expectations. Because our model is deterministic, this means that workers have perfect foresight.

Each sector of the DC uses a fixed input and a mobile input (labor) to produce a single good with decreasing returns to scale.[9] Superscripts on production variables denote the sector and superscripts on consumption variables denote the relevant economic agent.[10] Subscripts denote partial derivatives. $L^i(t), i = 1, 2,$ is the labor used by the ith sector at time t. Time is continuous. $\hat{L}$ is the DC's fixed labor endowment. This means that $L^1(t) + L^2(t) = \hat{L}$. Good 2 is the import competing and the polluting good. Let $\tau_e(t)$ denote the pre-existing tariff that protects sector 2. The government's environmental policy instrument is a pollution tax, $\tau_p(t)$, that is levied on the production of good 2.

Following Karp and Paul (1994) and Batabyal (1998), we use duality theory to model consumption and production decisions in the DC. The production function in the ith sector, $i = 1, 2$, is $f^i(L^i)$. The world price of good 2 is $p = p^2/p^1$, where $p^1 = 1$. Let $L^2 = L$, and let $L^1 = \hat{L} - L$. Finally, denote the two revenue functions by $R^1(1, \hat{L} - L)$ and $R^2(p + \tau_e - \tau_p, L)$, respectively.[11] As

---

9. Returns to scale is a property of the production function of a sector. In particular, suppose there is a n-fold increase in all inputs. If output increases n-fold then the production function exhibits constant returns to scale. If output does not increase n-fold then we have decreasing returns to scale. Finally, if there is more than a n-fold increase in output then we have increasing returns to scale. For additional details, see Varian (1992, pp. 14–17).

10. An economic agent is an economic actor or a player who forms an important part of the underlying analysis. In this chapter, there are three types of agents. These are workers in the traditional sector, workers in the modern sector, and a residual claimant who is typically either a capitalist or the government.

11. The maximum revenue obtainable from given inputs (or factors) at given output prices is denoted by a function called the revenue function. This explains why the revenue function depends on output prices and on input quantities. In our case, this means that the two revenue functions depend on the two output prices and on the three inputs. Further, because we have selected good 1 to be the numeraire, its price equals unity. This is why the first argument of $R^1(\cdot, \cdot)$ is 1. In the second sector, the import tariff has to be added and the pollution tax has to be subtracted from the output price. This explains why the first argument of $R^2(\cdot, \cdot)$ is $p + \tau_e - \tau_p$. The fixed factors in the two sectors are not of interest to us. What is of interest is the mobile factor, i.e., labor. This is why the second argument of both revenue functions is labor. The reader should

explained in footnote 11, $R_1^i(\cdot)$ and $R_2^i(\cdot)$ are the output supply of good i and the wage in sector i, respectively.

There is a continuum of identical workers[12] in each sector in the DC and a single capitalist is the residual claimant. All agents have homothetic preferences.[13] Then, following Dixit and Norman (1980, p. 326), the expenditure function of agent $j, j = 1, 2, 3$, is $\bar{E}(p + \tau_e, 1, +w^j) = U^j E(p + \tau_e)$, where $E(\cdot)$ is the unit expenditure function and U^j is agent j's real income. Our DC's national income is $U \equiv (\hat{L} - L)U^1 + LU^2 + U^3$. The superscript j stands for the representative worker in sector $j = 1, 2$, and $j = 3$ stands for the capitalist.

The private value of migration for a worker at time t is $m(t)$. That is, $m(t)$ denotes the discounted value of the wage differential between the high wage polluting sector and the low wage non-polluting sector. Formally, we have

$$m(t) = \int_t^\infty e^{-r(s-t)} \{R_2^2(\cdot) - R_2^1(\cdot)\} ds, \qquad (1)$$

where r is the discount rate. Using Leibnitz's rule for differentiating under an integral (see Kamien and Schwartz, 1991, p. 292), we can express Equation (1) as a differential equation. That equation is

$$\dot{m} = rm + R_2^1(\cdot) - R_2^2(\cdot). \qquad (2)$$

A worker will migrate to the high wage sector only when the private value of migration, $m(t)$, is at least as high as the private

---

note that we have suppressed the dependence of the revenue function on the fixed factors of production. Because of the nature of the maximization problem that yields the revenue function, the partial derivatives of the revenue function have economic meanings. In particular, "the partial derivatives ... of a revenue function with respect to the output prices yield the output supplies" (Dixit, 1976, p. 36). Similarly, the partial derivatives of the revenue function with respect to the input quantities yield the input prices. This explains why $R_1^i(\cdot)$ and $R_2^i(\cdot)$ denote the output supply of good i and the wage in sector i. Readers interested in additional details should consult Dixit (1976, pp. 36–37) and particularly Dixit and Norman (1980, Chapter 2).

12. Mathematically, the word "continuum" refers to the set of real numbers. What we mean here is that each sector of the DC is populated by a very large number of identical workers.

13. The implication here is that the preference functions can be expressed as monotonic transformations of functions that are homogeneous and of degree 1. What this means for our purpose is that the relevant expenditure functions can be written in a particular way. For more details, see Varian (1992, pp. 146–147).

cost of migration. Nevertheless, because workers do not account for pollution in their migration decisions, the social cost of migration is unequal to the private cost of migration. Let the social cost of migration be quadratic. Then $C(\dot{L}) = \alpha(\dot{L})^2, \alpha > 0$. Here, we are thinking of migration as the rate of change in the sector 2 labor stock. Because the average social cost of migration, $\alpha \dot{L}$, is smaller than the marginal social cost, $2\alpha \dot{L}$, in the absence of governmental intervention, migration for high wage employment in the polluting sector occurs too rapidly and increases environmental degradation.

To account for the fact that the private cost of migration is less than the social cost, suppose that workers base their migration decision on a fraction $\delta, 0 < \delta < 1$, of the marginal social cost $2\alpha \dot{L}$. This means that the migrating workers do not internalize the externality (the presence of pollution) arising in part from their decision to migrate. Let us now equate the private value of migration with the private cost of migration. This gives us the following equation for the dynamics of labor migration:

$$\dot{L} = \frac{m}{2\alpha\delta}. \quad (3)$$

Note that we are analyzing a trading DC. Moreover, we are disallowing the possibility of international borrowing. Consequently, in equilibrium, the deficit in the balance of trade must equal zero. Let us denote the deficit in the balance of trade by the function $D(U, L, m, \tau_e, \tau_p)$. This function depends on national income, labor in sector 2, the private value of the wage differential between the two sectors, the import tariff, and the pollution tax. The equilibrium condition that we just mentioned can be written compactly as $D(U, L, m, \tau_e, \tau_p) = 0$.[14] With full details, the same equilibrium condition reads as follows:

$$D(U, L, m, \tau_e, \tau_p) = UE(\cdot) + \frac{m^2}{4\alpha\delta^2} - R^1(\cdot) - R^2(\cdot)$$
$$- \tau_e[UE_1(\cdot) - R_1^1(\cdot)] - \tau_p R_1^2(\cdot) = 0. \quad (4)$$

---

14. This condition is like a budget constraint. For more on this type of constraint, see Obstfeld and Rogoff (1996).

The first term on the right hand side (RHS) of this balance of trade deficit equation refers to consumption expenditures. Equation (3) tells us that $C(\dot{L}) = m^2/4\alpha\delta^2$. Therefore, the second term on the RHS of Equation (4) denotes the social cost of pollution. The third and the fourth terms give the value of production. Finally, the fifth and the sixth terms denote the tariff and the tax revenues. We assume that these revenues are redistributed in lump sum fashion.

Our aim now is to study the DC government's optimal dynamic environmental policy under three assumptions about its ability to commit to a particular course of action. In the first case, the government commits to a tax trajectory or program for an infinite period of time. This infinite period of commitment should be interpreted as a case in which environmental protection is enshrined in the constitution. As noted in Batabyal (1998), if the DC in question were India, then this period would be 1976. This is because until 1976, environmental protection did not appear anywhere in the Indian constitution. Obviously, when environmental protection is enshrined in the constitution, it does not matter which government is in power because the constitution will have to be followed. In the second case, the DC government commits to its tax trajectory for a finite period of time. This finite period of commitment is more plausible and it should be thought of as the length of time during which a particular government is in office. Regrettably, in both these cases, the government's optimal tax policy is time inconsistent. To grasp this, consider the tax trajectory that the government announces at time $t = 0$. Time inconsistency means that at some time $\varepsilon > 0$, the government will depart from the trajectory it announced at $t = 0$. As a result, the government's announced policy at time $t = 0$ is not credible. This means that forward looking workers will not believe that the government will actually carry through with its initially announced policy. Therefore, this policy will fail to accomplish its intended objectives.

Since the credibility of government policy has been a salient issue in many DCs, *a priori*, it would seem necessary to study the implications of the DC government following a time consistent course of action. This is the third case that we study. In this scenario, the government commits to its tax policy for an infinitesimal period of time. In the limiting case in which the period of commitment approaches zero, the government's tax policy is time consistent. This completes the discussion of our theoretical framework. We now turn to the DC government's problem when it can commit to its tax policy for an infinite period of time.

3. The Infinite Commitment Case

In this case, the DC government makes a binding commitment and chooses its tax trajectory from time $t = 0$ to $t = \infty$, at $t = 0$. This is the government's open loop tax policy. The open loop pollution tax is a function of calender time only. Recall that workers have perfect foresight and that they are forward looking. Also recall that because of the presence of distortions, the DC economy is initially (at time $t = 0$) in disequilibrium.[15] As a result, the initial value of L, $L(0) = L_0$, is unequal to its steady state value in the polluting sector of the economy.

The decision to migrate is an investment decision. So, the private value of migration at any time t, $m(t)$, is determined by the current and the *future* values of the pollution tax. This means that the constraint in Equation (2) is a jump state constraint.[16] Formally, this means that the initial value of m, $m(0)$, is endogenous to the control problem being analyzed. In this setting, the DC government solves

$$\max_{U,\tau_p} \int_0^\infty e^{-rs} U ds, \qquad (5)$$

---

15. The reader may want to revisit the discussion of this equilibrium in the third paragraph of Section 2.

16. Many problems in economics are characterized by the existence of jump states. For instance, in monetary economics, the exchange rate is a jump state because it is affected by current interest rates and agents' expectations of the future money supply. For more on jump state constraints, see Karp and Newbery (1993) and Karp and Paul (1994).

subject to Equations (2)–(4), with initial condition $L(0) = L_0$. The current value Hamiltonian for this problem is

$$H = U - \lambda\left[UE + \frac{m^2}{4\alpha\delta^2} - R^1 - R^2 - \tau_e UE_1 + \tau_e R_1^2 - \tau_p R_1^2\right]$$

$$+ \sigma_1\left\{\frac{m}{2\alpha\delta}\right\} + \sigma_2\{rm + R_2^1 - R_2^2\}, \tag{6}$$

where λ is the Lagrange multiplier on constraint (4), and σ_1, σ_2 are the costate variables associated with constraints (3) and (2), respectively. The first-order necessary conditions are

$$\lambda = \frac{1}{E(p + \tau_e) - \tau_e E_1(p + \tau_e)}, \tag{7}$$

$$\lambda\{(\tau_e - \tau_p)R_{11}^2(\cdot)\} + \sigma_2 R_{21}^2(\cdot) = 0, \tag{8}$$

$$\dot{\sigma}_1 = r\sigma_1 + \sigma_2 h(\cdot) + \lambda\{d(\cdot) + (\tau_e - \tau_p)R_{12}^2(\cdot)\}, \tag{9}$$

and

$$\dot{\sigma}_2 = \frac{\lambda m}{2\alpha\delta^2} - \frac{\sigma_1}{2\alpha\delta}, \tag{10}$$

where $d(\cdot) \equiv R_2^1(\cdot) - R_2^2(\cdot)$ and $h(\cdot) \equiv R_{22}^1(\cdot) + R_{22}^2(\cdot)$. Put differently, $-d(\cdot)$ is the current private value of migration, and $h(\cdot)$ denotes the sum of the slopes of the marginal products of labor in the two sectors. Note that $h(\cdot) = \partial\{-d(\cdot)\}/\partial L < 0$.

Our primary interest lies in describing the optimal pollution tax trajectory, and in studying the dependence of this tax on the tariff $\tau_e(t)$. To this end, let us denote steady state values, i.e., the values of the relevant variables at time $t = \infty$ with the superscript S. By setting $dL/dt = 0$ in Equation (3), we get $m^S = 0$. Equation (2) gives $d^S(\cdot) = 0$. Equation (10) implies that $\sigma_1^S = 0$. From Equation (8) it follows that $\sigma_2^S = [-\lambda\{(\tau_e - \tau_p)R_{11}^2\}/R_{21}^2]^S$. From Equation (9), we get $\sigma_2^S = [-\lambda\{(\tau_e - \tau_p)R_{12}^2\}/h]^S$. Setting these last two expressions equal, we get $(\tau_e^S - \tau_p^S)\{R_{11}^2/R_{21}^2\}^S = (\tau_e^S - \tau_p^S)\{R_{12}^2/h\}^S$. Now note that $R_{11}^2 > 0$ and $h < 0$. Further, R_{21}^2 and R_{12}^2 will generally be positive. Using this information, we see that the coefficients multiplying

$(\tau_e^S - \tau_p^S)$ in the preceding equation have opposite signs. As such, it must be the case that $(\tau_e^S - \tau_p^S) = 0$. This tells us that $\tau_e^S = \tau_p^S$. From Equation (8), it follows that $\tau_p(t) = \tau_e(t) + \sigma_2(t)R_{21}^2/\lambda(t)R_{11}^2$. Because $m(0)$ is free, as Simaan and Cruz (1973) have noted, the correct boundary condition for σ_2 is $\sigma_2(0) = 0$. In other words, the DC government chooses its pollution tax trajectory so that the social shadow value (or imputed value) of m at the beginning of the program is zero. Using $\sigma_2(0) = 0$, we get $\tau_p(0) = \tau_e(0)$.

3.1. Discussion

Inspection of the expressions for $\tau_p(0)$, $\tau_p(t)$, and τ_p^S from the previous paragraph tells us that in an optimal program, the government's pollution tax depends on the existing tariff τ_e, in a straightforward manner. First, at the beginning and at the end of the program, the magnitude of the optimal pollution tax is equal to the magnitude of the existing tariff and both are positive. Second, in general $\sigma_2(t) > 0, R_{21}^2(\cdot) > 0, \lambda(t) > 0$, and $R_{11}^2(\cdot) > 0$. This means that $\{\sigma_2(t)R_{21}^2(\cdot)/\lambda(t)R_{11}^2(\cdot)\} > 0$. So the optimal pollution tax at an interior point in the program will generally be larger than the existing positive tariff. Putting these two pieces of information together, we conclude that in an optimal program, the government begins with a positive pollution tax that is equal in magnitude to the tariff, then raises this tax, and finally lowers this tax so that in the steady state the pollution tax and the tariff are once again equal in magnitude and positive. Note that there is no jump in policy at time $t = \infty$. The government simply lowers the pollution tax gradually to its optimal steady state value.

There are two distortions in our DC economy — the import tariff and pollution — and the government has available to it a single policy instrument, namely, the pollution tax. Standard welfare economics tells us that in general, for there to be an improvement in welfare, the number of policy instruments must equal the number of distortions. In our case, this means that the government will not be able to use environmental policy to raise welfare unambiguously.

The results of the previous paragraph should be interpreted in the context of this second best environment (see footnote 4) in which the DC government operates. At $t = 0$, we have $\tau_p(0) = \tau_e(0) > 0$. The positive tariff results in excess production of the good manufactured by the import competing sector. Consequently, here, the positive pollution tax simply offsets this excess production effect of the tariff. This tax is unable to simultaneously deal with the pollution distortion. As such, the pollution distortion remains unaddressed.

At $t = \infty$, once again we have $\tau_p^S = \tau_e^S > 0$. Now, the situation is different. In the steady state, all adjustments in the economy have taken place and there is no growth in pollution. Also, there is no rationale for migrating to the polluting sector because $m^S = 0$. In other words, there is no pollution externality and hence the partial internalization of this externality by the workers is not an issue. There is only one distortion in the steady state (the import tariff) and as in the $t = 0$ case discussed in the previous paragraph, the positive pollution tax offsets the excess production effect of the tariff. Note that in this steady state, and only in this steady state, environmental policy unambiguously improves welfare because it addresses the only distortion in the DC economy.

At any interior $t \in (0, \infty)$, both distortions are present in the DC economy. Also, both these distortions affect the output of the polluting good in the same way: they result in over-production. This is why in general we have $\tau_p(t) > \tau_p(0) > 0$ and $\tau_p(t) > \tau_p^S > 0$. The pollution tax at any interior point in the optimal program attempts to address both distortions in the economy; however, this single instrument does so only imperfectly. This explains why $\tau_p(t)$ is larger in magnitude than the two pollution taxes at the beginning and at the end of the government's program.

In this open loop case that we have been studying so far, there is no welfare loss to society from the government's inability to commit to its announced policy. This is because the open loop policy incorporates perfect or infinite commitment. Consequently, the case for doing nothing, i.e., setting a zero pollution tax, which potentially arises when the government cannot commit, is ruled out. In other words, intuitively, we expect the government's optimal environmental policy to be activist. From the analysis thus far, we see that this is indeed the case because in general $\tau_p(0), \tau_p^S$, and $\tau_p(t), \forall t \in (0, \infty)$, are all positive.

From the standpoint of policy credibility, if the DC government's open loop tax policy is believed by the migrating workers, then this policy will achieve its objectives. In particular, this tax will reduce output and employment in sector 2 and slow the rate of migration from the non-polluting sector 1 to the polluting sector 2. However, the government's objectives will *not* be met because this government will have an incentive to depart from the policy it announced at $t = 0$. To comprehend this, note that for any initial value of $L, L(0) \neq L^S$, from Simaan and Cruz (1973), the optimal initial shadow value of $m(t), \sigma_2(t)$, is zero. However, because $\delta < 1$, on the announced tax trajectory, $\sigma_2(t) \neq 0$. Consequently, at any time $\varepsilon > 0$, the government will want to depart from the tax trajectory it announced at $t = 0$ and announce a new trajectory. Put differently, the government's open loop tax policy is time inconsistent. This means that unless there is some mechanism by which the DC government can be bound to its initially announced pollution tax trajectory, this government will fail to achieve its environmental and employment objectives.

From a practical perspective, this case of perfect commitment is clearly farfetched because no government can realistically be expected to commit to its policy for an infinite period of time. Consequently, we now examine the case in which the DC government commits to its announced policy at the beginning of its tenure in office, for a finite period of time. This is the limited commitment case.

4. The Limited Commitment Case

Given that governments are in office for a finite period of time, the most sensible period of commitment equals the length of time during which a particular government is in office. Consequently, let us now analyze the limited commitment case in which the DC government commits to a policy for $T \in (0, \infty)$ time periods.

When the period of commitment is finite, the ensuing equilibrium is a function of the manner in which economic agents form their expectations. If migrating workers condition their expectations of future taxes on the history of taxes, then there will generally be multiple equilibria. To get around this problem, we shall limit our attention to smooth or differentiable Markov perfect equilibria.[17] In this context, Markov means that all agents have rational expectations but they condition their expectations on the payoff relevant state variable only. Put differently, the decision rules of the agents at any time t, depend only on the current value of the stock of labor (the state variable) at time t, and not on the manner in which the current stock of labor was attained. A prospect for an equilibrium is perfect if this prospect is an equilibrium for any subgame, i.e., for any level of the stock of labor. Specifically, whether or not some agents have departed from their equilibrium strategies in the past, the continuation of these strategies represents equilibrium behavior on the part of all the agents involved. From a practical perspective, this Markov assumption is of value because it makes the DC government's optimal program unresponsive to agent's mistakes.

Given this restriction of Markov perfection, we are now in a position to describe the equilibrium that emerges when the government commits to its tax policy for T periods. At time periods $0, T, 2T, \ldots$, consecutive governments choose their own tax policies. This means that at each $iT, i = 0, 1, 2, \ldots$, the ith government completes its tenure in office and a new government selects its tax policy for the next T time periods. At the end of T periods, each government bequeaths L_T, the current stock of labor, to its successor government. This government then conducts environmental policy for the next T periods, and so on.

---

17. Readers interested in a detailed discussion of the Markov perfect equilibrium concept should consult Fudenberg and Tirole (1991, Chapter 13).

With this interpretation of the limited commitment case, let $V(L)$ be the value of the government's program when its period of commitment is T periods and when the initial level of labor in the polluting sector is L. The government now solves

$$V(L) = \max_{\tau_p, U} \int_0^T e^{-rt} U dt + e^{-rT} V(L_T), \qquad (11)$$

subject to Equations (2)–(4). $V(L_T)$ is a bequest function that denotes the value of the stock of labor in sector 2 bequeathed by an arbitrary government to its successor. Note that problem (11) is the same as problem (5) in Section 3, with the exception that the government's period of commitment is now T and not infinity. This change will alter the boundary conditions at the horizon of the optimal program; however, the first-order necessary conditions themselves remain as in Equations (7)–(10).

As in Section 3, $m(0)$ is free. Hence, it is optimal to select the tax trajectory so that $\sigma_2(0) = 0$. Using this last condition in Equation (8), we get $\tau_p(0) = \tau_e(0)$. Once again, as in Section 3, $\tau_p(t) = \tau_e(t) + \sigma_2(t) R_{21}^2 / \lambda(t) R_{11}^2$. Finally, to determine $\tau_p(T)$, let $M(L)$ be the equilibrium current value of m that is determined by the solution to problem (11).[18] In our case, we can write $V(L) \equiv \bar{V}\{L, M(L)\}$, for some function $\bar{V}\{\cdot\}$. At the beginning of a specific time period $iT, i = 0, 1, 2, \ldots, \sigma_2(iT) = 0$. Further, the assumed smoothness or differentiability of the value function gives $\sigma_2 = \partial \bar{V}/\partial M$ (Karp and Paul, 1994, p. 1388; Batabyal, 1998, p. 15). This means that the social shadow value of M is equal to the marginal value of M in the bequest. Finally, the transversality condition for σ_2 is $\sigma_2(T) = \partial \bar{V}/\partial M = 0$. Using this last condition

---

18. The assumptions made in the second paragraph of this section and, in particular, this endogenous function of the state ($M(L)$) are features of the Markov perfect equilibrium that we are analyzing. Further, note that the open loop solution of Section 3 and the Markov perfect solution of this section are related in the sense that both solutions refer to an equilibrium of the underlying model. However, because specific assumptions need to be made to derive the Markov perfect equilibrium, this equilibrium is different from the open loop equilibrium of Section 3. The properties of this endogenous function of the state have been discussed in detail in Karp and Newbery (1993) and in Karp and Paul (1994). Consequently, we omit an elaborate discussion.

in Equation (8), the value of the optimal pollution tax at time $t = T$ is $\tau_p(T) = \tau_e(T)$.

4.1. Discussion

Comparing the three tax expressions in the previous paragraph with the corresponding tax expressions from Section 3, we see that a diminution in the length of the government's period of commitment results in no qualitative change in $\tau_p(0), \tau_p(t)$, or in $\tau_p(T)$. Consequently, in general, the analysis of Section 3.1 applies to this limited commitment scenario as well. However, there is one difference. In Section 3, in the steady state $m^S = 0$ and hence in this case there is a single distortion in the economy (the tariff) and the pollution tax clearly raises welfare. However, in the limited commitment case, in general, we will not have $m(T) = 0$. Consequently, in this case, in an optimal program, the government operates in a second best environment (see footnote 4) at all points in time. From a welfare perspective, this means that the DC may or may not be better off with the government's activist environmental policy.

Note the significant role played by the endogenous function of the state variable, $M(L)$. This function performs the role of an "expectations" function. When the DC government solves its maximization problem taking this expectations function as exogenous, the optimal program results in an initial value of $m, m(0)$, that satisfies $m(0) = M\{L(0)\}$. Put differently, in equilibrium, every agent's point expectations are satisfied. Further, this same optimal program results in a terminal value of m such that $\partial \bar{V}(\cdot)/\partial M = \sigma_2(T) = 0$. This means that at the horizon of the program, i.e., at time $t = T$, the shadow or imputed value of the state M, equals the marginal value of M in the bequest function $V(\cdot)$, and these two values equal zero.

Even though this limited commitment scenario is believable, the attendant Markov perfect equilibrium is time *inconsistent.* To see why, think of this Markov perfect case as one in which an infinite sequence of governments conducts environmental policy during a time period of length T. Denote the tenure of each government in this sequence by $\{iT\}_{i=0}^{\infty}$. When $T > 0$, each government behaves consistently at each i, but not within a period of length T. In other words, the DC government begins its term of office with the best of intentions, but some time later, it reneges on the policy it announced at the beginning of its term of office. As a result, forward looking agents will not believe that the government will actually carry through with its initially announced policy. From the standpoint of believability, this means that the government will fail to accomplish its policy objectives. In particular, even with an activist environmental policy, pollution and employment in sector 2 will not be reduced, and the government will not succeed in slowing the migration rate from the low wage traditional sector to the high wage polluting sector.

So far we have seen that the time inconsistency of the government's optimal environmental policy can prevent the DC government from attaining its environmental and employment goals. How can the time inconsistency of the government's optimal tax policy be eliminated? We now show how this can be done by studying a case in which the DC government commits to its environmental policy for an infinitesimal period of time. In this setting, we analyze the limiting Markov perfect equilibrium in which the government's period of commitment shrinks to zero.

5. The Infinitesimal Commitment Case

Intuitively, we expect the government's equilibrium pollution tax to be a function of three elements. The first element — the presence of pollution — generally calls for an activist policy that will correct for this negative externality. The second element — the government's inability to commit to its tax trajectory — would appear to favor a "do nothing" course of action. The third element — the presence of the tariff — encourages over-production of the import competing good; consequently, in general, this element also calls for an activist course of action. Given this situation, we now ask the following question: When the government's period of commitment is infinitesimal, is it ever optimal to set a zero pollution tax? In other words, is it possible for the "do nothing" course of action to dominate the activist course of action?

In our study of this infinitesimal case, we shall follow Karp and Paul (1994) and Batabyal (1998). We begin with a discrete stage formulation of the DC government's problem.[19] Denote the government's period of commitment and the length of each stage by ε. Further, suppose that all agents act at the beginning of each time period of length ε. Then, at time t, the government faces constraints (3) and (2). In discrete form, these two constraints are

$$L_t = \left\{\frac{m_t}{2\alpha\delta}\right\}\varepsilon + L_{t-\varepsilon} \qquad (12)$$

and

$$m_t = e^{-r\varepsilon}m_{t+\varepsilon} - d_t(\cdot)\varepsilon, \qquad (13)$$

where $d_t(\cdot) = R_2^1(\cdot) - R_2^2(\cdot)$. In Equation (12), $\{m_t/2\alpha\delta\}\varepsilon$ stands for the number of migrants in a period of length ε. Similarly, in Equation (13), $-d_t(\cdot)\varepsilon$ represents the value of the flow of the wage differential in a time period of length ε. We are now in a position to state the government's maximization problem.

At time t, with period of commitment ε, the DC government's dynamic programming problem is

$$V(L_{t-\varepsilon}) = \max_{U,\tau_p}[U - \lambda\{D(U, L, m, \tau_e, \tau_p)\}]\varepsilon + e^{-r\varepsilon}V(L_t), \qquad (14)$$

subject to Equations (12) and (13). Observe that the function $D(\cdot)$ represents the "balance of trade deficit" constraint described by Equation (4), that $m_{t+\varepsilon} = M(L_t)$, and that the government takes the function $M(\cdot)$ as exogenous. After some algebra, the first-order

---

19. For additional details on the underlying methodology, see Karp and Newbery (1993).

necessary condition to problem (14) w.r.t. τ_p can be written as

$$\left[\lambda\left\{(\tau_e - \tau_p)R_{11}^2 - \frac{\partial D}{\partial L_t} \cdot \frac{dL_t}{d\tau_p} - \frac{\partial D}{\partial m_t} \cdot \frac{dm_t}{d\tau_p}\right\}\right]\varepsilon + e^{-r\varepsilon}\frac{dV}{dL_t} \cdot \frac{dL_t}{d\tau_p} = 0. \tag{15}$$

In order to simplify Equation (15), let us differentiate Equations (12) and (13) totally. We get

$$\frac{dL_t}{d\tau_p} = \frac{\varepsilon}{2\alpha\delta} \cdot \frac{dm_t}{d\tau_p}, \tag{16}$$

and

$$\left\{-b_t(\cdot)\varepsilon - e^{-r\varepsilon}\frac{dM}{dL_t}\right\}\frac{dL_t}{d\tau_p} + \frac{dm_t}{d\tau_p} = -\left\{\frac{\partial d_t(\cdot)}{\partial \tau_p}\right\}\varepsilon. \tag{17}$$

Now substitute for $dL_t/d\tau_p$ from Equation (16) into Equation (17) and then simplify the resulting expression. This gives $dm_t/d\tau_p \sim O(\varepsilon)$. Similarly, substituting for $dm_t/d\tau_p$ from Equation (17) into Equation (16) and then simplifying the resulting expression yields $dL_t/d\tau_p \sim o(\varepsilon)$.[20] Finally, divide both sides of Equation (15) by ε, use the preceding two results about $dm_t/d\tau_p$ and $dL_t/d\tau_p$, and then let $\varepsilon \to 0$. The limiting first-order necessary condition is

$$\lambda(\tau_e - \tau_p)R_{11}^2 = 0. \tag{18}$$

5.1. *Discussion*

Equation (18) tells us that the limiting Markov perfect pollution tax $\tau_p = \tau_e$. We see that this limiting tax is also positive and equal in magnitude to the tariff. This result enables us to provide a clear answer to the question that was posed in the first paragraph of this section. In particular, even when the DC government's period of commitment is infinitesimal, it is not optimal for the government to set a zero pollution tax. Put differently, the activist course of action dominates the "do nothing" or passive course of action.

---

20. For additional details on this notation, see Karp and Paul (1994) and Ross (1996, p. 60).

The infinitesimal case that we are studying in this section requires the government to continuously revise its pollution tax. When this government revises its policy instrument continually, the ensuing policy is time consistent. This means that the government's environmental policy is credible. Temporarily, let us set the tariff aside and focus on the believability aspect of intertemporal environmental policy. As Karp and Newbery (1993) have noted, the payoff to an economic agent is monotonic in his period of commitment. Consequently, reducing the government's period of commitment can never make this government better off. With this remark and the previous discussion of policy efficacy in mind, let us rank the three policies in term's of the government's preference, and the policy's ability to achieve its objectives.

From the DC government's perspective, the most desirable policy is the open loop policy because this policy leads to the highest payoff for the government. The second best policy is the Markov perfect tax policy with a finite period of commitment. The least desirable policy is the limiting Markov perfect tax policy with an infinitesimal period of commitment. In contrast with this ranking, the ranking in terms of goal attainment is reversed. The limiting Markov perfect tax policy is credible. As such, this policy will be believed by the workers and hence this policy will be able to reduce pollution and slow migration to the polluting sector. The other two policies are not believable; hence they will fail to achieve the government's environmental and employment goals. This discussion highlights the DC government's dilemma. The policy which results in the highest payoff to the government is the one that is least desirable from a credibility perspective.

6. Conclusions

A dynamic version of the Ricardo–Viner model was used in this chapter to study a dualistic DC economy in which there is pollution and the import competing sector is protected with a tariff. We examined the conduct of dynamic environmental policy by the

DC government under three assumptions about this government's ability to commit to its announced policy. Four significant policy conclusions emerge.

First, our analysis shows that doing nothing, i.e., setting a zero pollution tax, is not an optimal course of action. In every case that we analyzed and no matter what the DC government's period of commitment, we showed that the optimal pollution tax is positive.

Second, the analysis of this chapter tells us that the time inconsistency of certain optimal programs may prevent the government from attaining its environmental and employment objectives. Our analysis demonstrated that as long as the private cost of migration is less than the social cost of migration, i.e., as long as $\delta < 1$, the limiting Markov perfect tax policy is the only believable environmental policy. Further, we showed that when the import competing sector is protected with a tariff, in general, the government cannot use environmental policy to unambiguously raise welfare in the DC. The conduct of environmental policy raises welfare unambiguously only in the steady state. As discussed in Section 3, this is because in the steady state, there is only a single distortion in the DC economy.

Third, from a policy believability perspective, our analysis points to the implausibility of time inconsistent, particularly open loop policies. Such policies will not be believed by forward looking agents with rational expectations. Hence, these agents will successfully thwart the DC government's policy objectives. In contrast to this, the limiting Markov perfect pollution tax policy is time consistent. In this case, the equilibrium is delineated by an endogenous function of the state variable and the government continuously revises its tax trajectory. Continuous revision implies credibility and this in turn means that the government's environmental policy will achieve its intended objectives.

Fourth, there is a tradeoff between policy payoff and policy believability. Credible policies yield a lower payoff than do incredible policies. This remark provides a likely explanation as to why many DC governments are loath to use time consistent policies that involve continuous policy revision.

The analysis contained in this chapter can be extended in a number of directions. In what follows, we suggest two possible extensions. First, one can drop the small country assumption and analyze environmental policy in a DC whose actions affect world prices. Second, one can analyze environmental policy in a setting in which the decision to protect the import competing and the polluting sector is endogenous to the DC government, i.e., this decision is determined by the underlying model. Studies which incorporate these aspects of the problem into the analysis will provide richer accounts of the nexuses between protection, time consistency, and dynamic environmental policy in DCs.

References

Batabyal, A.A. (1995). Development, Trade, and the Environment: Which Way Now? *Ecological Economics* 13:83–88.

Batabyal, A.A. (1998). Environmental Policy in Developing Countries: A Dynamic Analysis. *Review of Development Economics* 2:293–304.

Bhalla, A.S. (1992). *Environment, Employment, and Development.* Geneva, Switzerland: International Labor Office.

Bonetti, S. and FitzRoy, F. (1999). Environmental Tax Reform and Government Expenditure. *Environmental and Resource Economics* 13:289–308.

Christainsen, G.B. and Tietenberg, T.H. (1985). Distributional and Macroeconomic Aspects of Environmental Policy. *In* A.V. Kneese and J.L. Sweeney (eds.), *Handbook of Natural Resource and Energy Economics*, Vol. 1. Amsterdam, The Netherlands: Elsevier.

Dixit, A.K. (1976). *The Theory of Equilibrium Growth.* London, UK: Oxford University Press.

Dixit, A.K. and Norman, V. (1980). *Theory of International Trade.* Cambridge, UK: Cambridge University Press.

Fudenberg, D. and Tirole, J. (1991). *Game Theory.* Cambridge, MA: MIT Press.

Kamien, M.I. and Schwartz, N.L. (1991). *Dynamic Optimization*, 2nd edn. Amsterdam, The Netherlands: North-Holland.

Karp, L. and Newbery, D.M. (1993). Intertemporal Consistency Issues in Depletable Resources. *In* A.V. Kneese and J.L. Sweeney (eds.), *Handbook of Natural Resource and Energy Economics*, Volume 3. Amsterdam, The Netherlands: Elsevier.

Karp, L. and Paul, T. (1994). Phasing in and Phasing Out Protectionism with Costly Adjustment of Labor. *Economic Journal* 104:1379–1392.

Krugman, P.R. and Obstfeld, M. (2000). *International Economics*, 5th edn. Reading, MA: Addison-Wesley.

Lekakis, J.N. (1991). Employment Effects of Environmental Policies in Greece. *Environment and Planning A* 23:1627–1637.

Mehmet, O. (1995). Employment Creation and Green Development Strategy. *Ecological Economics* 15:11–19.

Miller, M. (1995). *The Third World in Global Environmental Politics*. Boulder, CO: Lynne Reinner Publishers.

Mussa, M. (1982). Government Policy and the Adjustment Process. *In* J.N. Bhagwati (ed.), *Import Competition and Response*, Chicago, IL: University of Chicago Press.

Obstfeld, M. and Rogoff, K. (1996). *Foundations of International Macroeconomics*. Cambridge, MA: MIT Press.

Renner, M. (1992). Jobs in a Sustainable Economy. *Worldwatch Paper* No. 104, Washington, DC.

Ross, S.M. (1996). *Stochastic Processes*, 2nd edn. New York: John Wiley and Sons.

Simaan, M. and Cruz, J.B. (1973). Additional Aspects of the Stackelberg Strategy in Non-Zero Sum Games. *Journal of Optimization Theory and Applications* 11:613–626.

Varian, H.R. (1992). *Microeconomic Analysis*, 3rd edn. New York: W.W. Norton and Company.

Chapter 12

ASPECTS OF THE THEORY OF ENVIRONMENTAL POLICY IN DEVELOPING COUNTRIES

With Hamid Beladi

We study two issues relating to the conduct of environmental policy in developing countries (DCs). First, when faced with a self-financing constraint, should an environmental authority (EA) raise/lower pollution taxes over time or should it run a deficit/surplus? Second, given recent findings about the dynamic inconsistency of optimal environmental policy, should an EA make its preferences about the relative benefits of environmental protection versus production public, or should it keep its preferences private? Our analysis reveals that when faced with a self-financing constraint, it is optimal for the EA to run a deficit/surplus. Second, social losses are lower when this EA keeps its preferences private.

1. Introduction

In contemporary times, the connections between the environment and development have come to dominate academic and public debate in most parts of the world. Three principal issues have been articulated by scholars working in this area. First, Bhalla (1992), Renner (1992), and Mehmet (1995) have made the case that it is important for developing countries (DCs) to implement policies that generate employment. Second, Goldin and Winters (1995), Faucheux *et al.* (1996), and more generally the sizeable literature on sustainable development[1] have stressed the need for instituting

---

1. For more on this literature, the reader should consult Atkinson *et al.* (1997), Farmer and Randall (1997), Pezzey (1997), and Heal (1998).

policies that protect the environment for the present and the future generations. Third, Batabyal (1998), Batabyal and Beladi (2002), and Lee and Batabyal (2002) have pointed out that under certain circumstances, employment creation and environmental protection are competing goals. What this means is that although DCs may begin the process of implementing environmental policies, over time, their commitment to such policies is likely to wane.

The purpose of this chapter is to study two aspects of the above-mentioned third issue. We analyze these two aspects by developing an alternate theoretical framework from that employed in Batabyal (1998), Batabyal and Beladi (2002), and Lee and Batabyal (2002). One of the key findings of these three papers is that in a dynamic setting, it is generally optimal for an environmental authority (EA) to *alter* the magnitude of pollution taxes over time. However, in many DCs, once pollution taxes have been set, from a political perspective, it is difficult to change — and in particular to raise — them.[2] Further, on account of litigation,[3] the need to grant subsidies,[4] and other reasons, EAs in many DCs incur substantial expenses in performing their regulatory duties.[5] Indeed, as Sinkule and Ortolano (1995, p. 143) have noted, in China, EAs see "their ability to pursue pollution control as directly linked to their own financial strength ..." This discussion raises a hitherto unanswered question about the nature of dynamic environmental policy in the presence of a (possibly) binding financial or budget constraint.[6] Consequently, we shall analyze the following question: When faced with a self-financing or budget constraint, is it still optimal for an EA to alter the trajectory

---

2. For more on this in the case of India, see Dwivedi (1997, pp. 103, 203).

3. Environmental litigation in China is discussed in Sinkule and Ortalano (1995), and Dwivedi (1997) and Mehta *et al.* (1997) discuss environmental litigation in India.

4. See Sinkule and Ortalano (1995, pp. 131–133) for subsidies in China and Dwivedi (1997) for subsidies in India.

5. Even when a legal proceeding has been instituted against non-compliant polluting firms, there is no assurance that this legal proceeding will result in success. For instance, in India, the prosecution success rate in 1992 in air and water pollution cases was 65% and 62%, respectively (Mehta *et al.*, 1997, pp. 23–24).

6. For more on the practical effects of budget constraints on the activities of EAs in China and India, see Sinkule and Ortalano (1995, p. 29) and Dwivedi (1997, pp. 124–125).

of pollution taxes over time? Or, depending on the actual expenses incurred, does it make more sense to run deficits/surpluses? This is the first question that is analyzed in this chapter.

The next question that we analyze is motivated by the findings in Batabyal (1998), Batabyal and Beladi (2002), and Lee and Batabyal (2002). These papers have shown that dynamic environmental policy in DCs is typically marked by an inability of the appropriate EA to commit to its announced course of action. In other words, the announced policy is *dynamically inconsistent*. Now given the dynamic inconsistency of optimal environmental policy, and the scant attention that this issue has received in the literature, one can ask what connection there exists between an EA's preferences and credible environmental policy. Specifically, should an EA make its preferences about the relative benefits of environmental protection versus production of the polluting good public, or should it keep its preferences private? This is the second question that we analyze.

The rest of this chapter is organized as follows: Section 2 analyzes a model of environmental policy in which the EA is constrained by the presence of a self-financing (budget) constraint. Section 3 studies a model of credibility in the conduct of environmental policy. Section 4 concludes and offers suggestions for future research.

2. Environmental Policy with a Self-Financing Constraint

2.1. *Preliminaries*

This section's model is adapted from Barro (1979). As in Batabyal (1998), we shall focus on a small, open, infinite horizon DC whose economy is dualistic. One sector is the traditional sector in which there is no pollution. The second sector is the modern sector in which production causes pollution. In the rest of this chapter, our attention will be on this polluting sector. Let q_t denote the output of the polluting sector at time t and let τ_t denote the tax levied by the EA on the production of this polluting good. Time is discrete and the price of the polluting good is normalized to unity. As a result of the imposition of this tax, the "socially correct" output of the polluting sector at time t is not q_t, but $q_t - a\tau_t^2/2$, where $a > 0$ is a parameter.

The representative consumer of this polluting good has a subjective time preference factor β. The interest rate is r. We suppose that $\beta = 1/(1 + r)$. Denote this consumer's period t consumption by c_t, and his utility from consumption c_t by $u(c_t)$.[7]

As indicated in Section 1, in the course of performing its regulatory functions, the EA necessarily incurs expenses in every time period. We model the necessity of this expenditure by letting the amount spent in each time period, e_t, be exogenous. The national government in the DC funds the EA — g_t per time period — for the discharge of its regulatory functions. From the EA's perspective, this allocation of funds is also exogenous.[8] We can now write this EA's self-financing (budget) constraint as

$$\sum_{s=t}^{\infty} \left[\frac{1}{1+r}\right]^{s-t} (\tau_s + g_s - e_s) = 0. \quad (1)$$

In addition to its environmental functions, the EA maximizes the representative consumer's lifetime utility. In other words, the EA chooses the time path of private consumption and pollution taxes to solve

$$\max_{\{c_s, \tau_s\}} \sum_{s=t}^{\infty} \left[\frac{1}{1+r}\right]^{s-t} u(c_s). \quad (2)$$

There are two other constraints on the EA's problem. The first is the polluting sector's budget constraint. This constraint is

$$\sum_{s=t}^{\infty} \left[\frac{1}{1+r}\right]^{s-t} \left[q_s - \frac{a\tau_s^2}{2} - \tau_s - c_s\right] = 0. \quad (3)$$

---

7. Note that consumption may be direct or indirect. For instance, in India, paper and sugar production are highly polluting activities (see Mehta *et al.*, 1997, pp. 107–108). It is clear that in the case of sugar, consumption is direct. However, with regard to paper, consumption may be direct or indirect.

8. The reader will note that this rules out the possibility of the EA lobbying for additional funds.

The second constraint arises from the nature of the EA's optimization problem. Because the subjective time preference factor equals the market discount factor ($\beta = 1/(1+r)$), the representative consumer's Euler equation for consumption in any two time periods s and $s+1$ is[9]

$$u'(c_s) = u'(c_{s+1}), \qquad (4)$$

where the prime denotes a derivative.

2.2. Analysis and Results

The EA maximizes the objective in Equation (2) subject to the constraints given in Equations (1), (3), and (4).[10] Let γ and δ be the Lagrange multipliers on constraints (1) and (3), respectively. Equation (4) tells us that the desired level of intertemporal consumption is constant at some level, say, $\hat{c}$. Taking this into account, the Lagrangian to problem (2) is

$$L = \frac{1+r}{r}[u(\hat{c}) - \delta \hat{c}]$$

$$- \sum_{s=t}^{\infty} \left[\frac{1}{1+r}\right]^{s-t} \left[\gamma(\tau_s + g_s - e_s) + \delta\left(q_s - \frac{\alpha \tau_s^2}{2} - \tau_s\right)\right]. \quad (5)$$

Differentiating Equation (5) w.r.t. $\hat{c}$ and τ_s, $\forall s \geq t$, we get the relevant first-order conditions. These are

$$u'(\hat{c}) = \delta, \qquad (6)$$

and

$$\gamma - u'(\hat{c}) = u'(\hat{c})\alpha \tau_s. \qquad (7)$$

Equation (6) tells us that at the optimum, the marginal utility of consumption equals the shadow value of the polluting sector's resources. Equation (7) is more instructive. This equation says that

---

9. For more on this, see Obstfeld and Rogoff (1996, p. 3).
10. Note that because both budget constraints bind in equilibrium, without loss of generality, we can express these two constraints as equalities.

at the optimum, there is a wedge between the shadow value of the EA's resources and the private value of consumption. This wedge (the LHS) equals the marginal deadweight loss of the pollution tax measured in terms of the representative consumer's utility.

It is important to note that both the shadow value of the EA's resources (γ) and the private value of consumption ($u'(\hat{c})$) are constant over time. This means that in our model, optimal pollution taxes are also constant over time. Put differently, like the representative consumer — who finds it optimal to smooth consumption over time — the EA also finds it optimal to smooth pollution taxes over time. The constant pollution tax can be computed from Equation (7). We get

$$\hat{\tau} = \frac{\gamma - u'(\hat{c})}{au(\hat{c})}, \qquad (8)$$

where γ satisfies

$$\gamma = u'(\hat{c}) + \left[\frac{r}{1+r}\right] u'(\hat{c}) \sum_{s=t}^{\infty} \left[\frac{1}{1+r}\right]^{s-t} a\{e_s - g_s\}. \qquad (9)$$

Equation (9) tells us that the EA's shadow value of resources (the LHS) equals the private value of consumption plus a weighted average of present and future marginal consumption costs arising from the difference between the necessary and exogenous stream of EA expenditures and government allotted funds. Put differently, at the optimum, there is a wedge between the shadow value of the EA's resources and the private value of consumption. This discussion highlights the fact that the EA's budget constraint has a very real effect on the optimal values of all the endogenous variables and in particular on the values of the optimal pollution taxes.

Because there is an optimal level of the pollution taxes, we can infer the following: Depending on the magnitude of the exogenous expenditures e_t, it will be optimal for the EA to run either a deficit or a surplus. We now summarize our main conclusions in the following proposition.

Proposition 1. *When faced with a self-financing constraint, the EA should set a constant pollution tax over time. When its expenditures, e_t, are unusually high, it will be optimal for the EA to run a deficit. Similarly, when its expenditures are unusually low, the EA should run a surplus.*

This concludes Section 2. We now study a model of credibility in the conduct of environmental policy.

3. Credibility and Environmental Policy

3.1. *Preliminaries*

One of the tasks of the EA of Section 2 is to set appropriate pollution taxes τ_t. In turn, these pollution taxes directly affect q_t, the production of the polluting good, and indirectly affect pollution x_t. In order to work with pollution directly, we suppose that the functional relationships between x_t, q_t, and τ_t are strictly monotonic. Formally, this will enable us to treat pollution and not pollution taxes as the EA's control variable. More informally, this will permit us to think of the EA as a "command-and-control" entity that sets pollution levels directly. Moreover, because we want to work with a loss function, it will be helpful to think of the EA as an entity that sets pollution levels (the bad) directly.

To reiterate, we shall think of x_t as the period t pollution level that is set by the EA in our DC. Let x_t^e denote the polluting sector's period $t - 1$ expectation of what pollution will be in period t. Assuming that all agents in our DC have rational expectations, we get $x_t^e \equiv E_{t-1} x_t$, where $E[\cdot]$ is the expectation operator. There will generally be some discrepancy between the EA's targeted output level of the polluting good[11] and the actual output level. To this end, let $w > 0$ denote the positive wedge between these two output levels. We suppose that the EA's preferences over pollution

---

11. As discussed in the previous paragraph, x_t, q_t, and τ_t are functionally related. Consequently, targeting pollution directly has the effect of targeting output indirectly.

and the production of the polluting good can be described by a loss function with the following form[12]

$$L_t = \frac{1}{2}x_t^2 - \theta_t(x_t - x_t^e - w). \tag{10}$$

In Equation (10), θ_t is a random variable with a two-point support. That is, $\theta_t = 0$ with probability p and $\theta_t = \Theta > 0$ with probability $q = (1 - p)$. The first term in the loss function in Equation (10) reflects a concern for pollution and the second term reflects a concern for the output of the polluting good. It is important to comprehend the meaning of the random variable θ_t. This random variable captures the effect of shocks that alter the relative benefits of pollution versus production. It can also be interpreted as the EA's "type." To see this, observe that $\theta_t = 0 \Rightarrow L_t = (1/2)x_t^2$. This corresponds to the case of a "green" or "environmentalist" EA that does not care about production and is concerned solely with pollution. On the other hand, when $\theta_t = \Theta$, we have a "conventional" EA that cares about both pollution and the production of the polluting good. From a political standpoint, a green EA is more likely with a liberal government in power and a conventional EA is more likely with a conservative regime in power.

3.2. Analysis and Results

Recall that we are interested in examining the link between the EA's preferences and credible environmental policy. Specifically, what we want to know is this: Should the EA make its preferences — knowledge of θ_t — about the relative benefits of environmental protection versus production public, or should it keep its preferences private? We shall answer this question in a series of steps.

Temporarily, let us suppose that the EA and the polluting sector interact only once. In this case, we want to compute the equilibrium

---

[12] This kind of loss function has been used in the monetary economics literature by Barro and Gordon (1983), Backus and Driffil (1985), and others. For a good account of dynamic consistency issues in monetary economics, see Obstfeld and Rogoff (1996, pp. 634–658).

levels of x_t and x_t^e in the one-shot game between the EA and the polluting sector. This is the *discretionary* case. To compute the equilibrium x_t, we shall take x_t^e as given, and differentiate Equation (10) w.r.t. x_t. The first-order necessary condition is $x_t - \theta_t = 0$, which tells us that in equilibrium

$$x_t = \theta_t. \tag{11}$$

This says that in equilibrium, the optimal level of pollution equals the EA's type. To compute the equilibrium value of x_t^e, we shall use the definition of x_t^e, Equation (11), and the assumption that all agents in the DC have rational expectations. This gives

$$x_t^e \equiv E_{t-1} x_t = E_{t-1} \theta_t = p \cdot 0 + q \cdot \Theta = q\Theta. \tag{12}$$

In other words, the equilibrium expected level of pollution equals the expected value of the random variable denoting the EA's type. What is the expected loss to the EA in the discretionary one-shot game equilibrium? To answer this, we use Equations (10), (11), and (12). We get

$$E_{t-1} L_t^D = E_{t-1}\left[\theta_t q\Theta + \theta_t w - \frac{\theta_t^2}{2}\right]. \tag{13}$$

Because $E_{t-1}[\theta_t^2] = q\Theta^2$, Equation (13) can be simplified. This simplification yields

$$E_{t-1} L_t^D = (q\Theta)^2 + \frac{q\Theta^2}{2} w - \frac{q\theta^2}{2}. \tag{14}$$

Having studied the one-shot discretionary game equilibrium and the associated expected loss to the EA, let us now focus on the notion of commitment.[13] We first want to compute the expected social loss from commitment $E_{t-1} L_t^C$. We will then compare this loss with Equation (14) to determine whether the expected loss to society under commitment is bigger or smaller than the expected loss with discretionary environmental policy. To compute $E_{t-1} L_t^C$, suppose that the EA can commit to a best response function for $x_t(\theta_t)$.

---

13. For an alternate perspective on commitment in environmental policy in DCs, see Batabyal (1998), Batabyal and Beladi (2002), and Lee and Batabyal (2002).

Let us compute the EA's optimal best response function given that $x_t^e = 0$. In other words, we want to compute the EA's optimal best response function given that the polluting sector believes that the EA is "green." Formally, this constraint can be written as

$$p \cdot x_t(0) + q \cdot x_t(\Theta) = 0. \tag{15}$$

The EA's objective function is

$$E_{t-1}L_t^C = p\left[\frac{\{x_t(0)\}^2}{2}\right] + q\left[\Theta w + \frac{\{x_t(\Theta)\}^2}{2} - \Theta x_t(\Theta)\right]. \tag{16}$$

Now incorporating the constraint (Equation (15)) into Equation (16), we can write the EA's problem as an unconstrained minimization problem. The EA solves

$$\min_{\{x_t(\Theta)\}} E_{t-1}L_t^C = \{x_t(\Theta)\}^2\left[\frac{q^2}{2p} + \frac{q}{2}\right] + q\Theta w - q\Theta x_t(\Theta). \tag{17}$$

The first-order necessary condition to this problem is

$$x_t(\Theta)\left[1 + \frac{q}{p}\right] = \Theta. \tag{18}$$

Equation (18) and further simplification give us the optimal values of the two pollution levels. We get $x_t(\Theta) = p\Theta$ and $x_t(0) = -q\Theta$. Substituting these two values into Equation (16) gives us an equation for the optimized value of the expected social loss from commitment. That equation is

$$E_{t-1}L_t^C = q\Theta w - \frac{pq\Theta^2}{2}. \tag{19}$$

We can now determine whether society is better or worse off with the EA committing to environmental policy. The relevant equations to compare are Equations (14) and (19). Inspection of these two equations and some algebra gives us the required result. We state this result as in the following proposition.

Proposition 2. $\forall p \in [0, 1), E_{t-1}L_t^C < E_{t-1}L_t^D$. When $p = 1$, $E_{t-1}L_t^C = E_{t-1}L_t^D$.

Proposition 2 tells us that in the general case, environmental policy with commitment results in lower social losses than does environmental policy with discretion. In other words, society is better off when the EA is committed to environmental policy. As one would expect, when there is no uncertainty about the EA's type, i.e., when $p = 1$, committed and discretionary environmental policy result in the same loss to society.

We are finally in a position to answer the question posed at the beginning of this subsection, i.e., should the EA make its preferences about the relative benefits of environmental protection versus production public, or should it keep its preferences private? Let us first analyze the case in which the EA reveals θ_t to the polluting sector. Suppose that the EA reveals θ_t on date $t = 1$, *before* x_t^e has been set by the polluting sector. In this case $x_t^e = \theta_t$. We will now compute the expected social loss from revelation on date $t - 2$.[14] Substitute $x_t^e = \theta_t$ and Equation (11) into Equation (10), the EA's loss function, and then take expectations. We get

$$E_{t-2} L_t^R = q\Theta w + \frac{1}{2} q\Theta^2, \qquad (20)$$

where the superscript R denotes revelation.

When the EA keeps the true value of θ_t secret, the polluting sector forms its expectations about pollution in accordance with Equation (12). In other words, $x_t^e = q\Theta$. Now substituting Equations (11) and (12) into Equation (10), the EA's loss function, and then taking expectations, we get an equation for the expected social loss from secrecy. That equation is

$$E_{t-2} L_t^S = q\Theta w + (q\Theta)^2 - \frac{1}{2} q\Theta^2. \qquad (21)$$

The superscript S in Equation (21) denotes secrecy. Now a comparison of Equations (20) and (21) and some algebra gives us the answer that we seek. We state this answer in the following proposition.

---

14. Note that it makes sense to perform this computation because on date $t - 2$, the EA itself does not know the true value of θ_t.

Proposition 3. $\forall q \in [0,1), E_{t-2}L_t^S < E_{t-2}L_t^R$. *When* $q = 1$, $E_{t-2}L_t^S = E_{t-2}L_t^R$.

Proposition 3 tells us that in the general case, the EA will prefer to precommit itself to *not* reveal θ_t before the polluting sector sets x_t^e. In the extreme case in which there is no uncertainty about the EA's type, the question of revelation versus secrecy is uninteresting because both actions result in identical social losses.

A salient implication of Proposition 3 is that the EA will actually prefer a system that mandates secrecy about its true preferences regarding the relative benefits of environmental protection versus production of the polluting good. In his book on environmental policy in India, Dwivedi (1997, p. 104) has noted that many environmental laws "confer enormous discretionary powers on administrative authorities." Propositions 2 and 3 together tell us that in general, this is *not* a good idea. In particular, our analysis shows that from the DC's perspective, it is better to have an EA that displays *commitment* to its environmental policy so that the polluting "industries know what to expect [and] how far to go with respect to changing their production processes …" (Dwivedi, 1997, p. 216).

4. Conclusions

In this chapter we analyzed two hitherto unstudied questions about the conduct of environmental policy in DCs. First, in Section 2, we established the proposition that when faced with a self-financing constraint, an EA should set a *constant* pollution tax over time. In particular, this means that when its expenditures are unusually high, it is optimal for this EA to run a deficit. Similarly, when its expenditures are unusually low, the EA should run a surplus.

Next, in Section 3, we analyzed the notion of credibility in environmental policy. We first demonstrated that environmental policy with commitment results in lower social losses than does discretionary environmental policy. Recently, in the context of India, Dwivedi (1997, p. 208) has argued that one way to improve environmental policy would be to increase the public's *awareness* of the different aspects of environmental regulation. In contrast, our analysis shows that society is better off when an EA keeps its preferences about the relative benefits of environmental protection versus production private. This tells us that a certain amount of secrecy in the conduct of environmental policy is a good thing.

The analysis of this chapter can be extended in a number of different directions. In what follows, we suggest two possible extensions. First, one could expatiate upon the model of Section 2 by studying the properties of the EA's optimal tax policy when its expenditures are endogenous. Second, with regard to the model of Section 3, it would be useful to study what effects more general loss functions have on the result that it is optimal for the EA to keep its preferences about the relative benefits of environmental protection versus production secret. Studies which incorporate these aspects of the problem into the analysis will provide richer accounts of the theory of environmental policy in DCs.

References

Atkinson, G., Dubourg, R., Hamilton, K., Munasinghe, M., Pearce, D. and Young, C. (1997). *Measuring Sustainable Development: Macroeconomics and the Environment*. Cheltenham, UK: Edward Elgar.

Backus, D. and Driffil, J. (1985). Rational Expectations and Policy Credibility Following a Change in Regime. *Review of Economic Studies* 52:211–221.

Barro, R.J. (1979). On the Determination of the Public Debt. *Journal of Political Economy* 87:940–971.

Barro, R.J. and Gordon, D.B. (1983). Rules, Discretion, and Reputation in a Model of Monetary Policy. *Journal of Monetary Economics* 12:101–121.

Batabyal, A.A. (1998). Environmental Policy in Developing Countries: A Dynamic Analysis. *Review of Development Economics* 2:293–304.

Batabyal, A.A. and Beladi, H. (2002). A Dynamic Analysis of Protection and Environmental Policy in a Small Trading Developing Country. *European Journal of Operational Research* 143:197–209.

Bhalla, A.S. (1992). *Environment, Employment, and Development*. Geneva, Switzerland: ILO.

Dwivedi, O.P. (1997). *India's Environmental Policies, Programmes, and Stewardship*. New York: St. Martin's Press.

Farmer, M.C. and Randall, A. (1997). Policies for Sustainability: Lessons from an Overlapping Generations Model. *Land Economics* 73:608–622.

Faucheux, S., Pearce, D. and Proops, J. (eds.) (1996). *Models of Sustainable Development*. Cheltenham, UK: Edward Elgar.

Goldin, I. and Winters, L.A. (eds.) (1995). *The Economics of Sustainable Development*. Cambridge, UK: Cambridge University Press.

Heal, G. (1998). *Valuing the Future: Economic Theory and Sustainability*. New York: Columbia University Press.

Lee, D.M. and Batabyal, A.A. (2002). Dynamic Environmental Policy in Developing Countries with a Dual Economy. *International Review of Economics and Finance* 11:191–206.

Mehmet, O. (1995). Employment Creation and Green Development Strategy. *Ecological Economics* 15:11–19.

Mehta, S., Mundle, S. and Sankar, U. (1997). *Controlling Pollution: Incentives and Regulations*. New Delhi, India: Sage Publications.

Obstfeld, M. and Rogoff, K. (1996). *Foundations of International Macroeconomics*. Cambridge, MA: MIT Press.

Pezzey, J.C.V. (1997). Sustainability Constraints versus "Optimality" versus Intertemporal Concern, and Axioms versus Data. *Land Economics* 73: 448–466.

Renner, M. (1992). Jobs in a Sustainable Economy. *Worldwatch Paper* No. 104. Washington, DC.

Sinkule, B.J. and Ortolano, L. (1995). *Implementing Environmental Policy in China*. Westport, CT: Praeger.

Chapter 13

PUBLIC VERSUS PERSONAL WELFARE: AN ASPECT OF ENVIRONMENTAL POLICYMAKING IN DEVELOPING COUNTRIES

In this chapter, we shed light on the nature of the interaction between an environmental authority (EA) and the polluting sector in a developing country (DC) when there is uncertainty about the relative weight that this EA places on public versus its own welfare. Within the context of this general issue, we answer three specific questions for any arbitrary time period t. First, we determine the expected level of pollution as well as the actual pollution in the polluting sector. Second, we compute the mean social loss arising in part from the uncertainty about the relative weight that the EA places on public versus its own welfare. Finally, we solve for the optimal value of the parameter which measures the relative weight the EA places on public versus its own welfare.

1. Introduction

Since the publication of the now prominent Brundtland Report (Brundtland, 1987), the environment has loomed large in virtually every discussion of what it means for the process of economic development to be sustainable. Although this notion of "sustainable development" now quite often means different things to different individuals, there is little debate on the basic point that the process of making economic development in the world's low income countries sustainable is fundamentally all about environmental protection.[1]

As Dwivedi and Khator (1995), Jan (1995), and Stoett (1995) have pointed out, in recent times, many developing countries (DCs) have adopted a number of measures to protect their environmental resources. However, because stringent environmental measures often inflict "pain" on certain sectors of a DC's economy, there is some concern among researchers and observers about the ability of DC governments to carry through with meaningful environmental policies. Put a little differently, because environmental protection and employment creation are often competing objectives, the worry is that although DCs may initiate the process of instituting and implementing environmental policies, over time, their faithfulness to such policies is likely to diminish.[2]

Recently, Batabyal and Beladi (2002b) have studied some of these issues concerning the conduct of environmental policy in DCs. Specifically, they pose and answer the following two questions. First, when faced with a self-financing constraint, should an environmental authority (EA) raise/lower pollution taxes over time or should it run a deficit/surplus? Second, should an EA make its preferences about the relative benefits of environmental protection versus production public, or should it keep its preferences private? Batabyal and Beladi (2002b) show that when faced with a self-financing constraint, it is optimal for the EA to run a deficit/surplus. Second, social losses are lower when this EA keeps its preferences private.

Despite the presence of these useful findings in the extant literature, a question that has not received adequate theoretical attention in the literature concerns the nature of the interaction between an EA and the polluting sector in a DC when there is uncertainty about the relative weight that this EA places on public versus its own

---

1. There is now a vast literature on the topic of sustainable development. For more on this literature, the reader should consult Atkinson *et al.* (1997), Dwivedi (1997), Farmer and Randall (1997), Pezzey (1997), Heal (1998), Munasinghe (2007), and Stern (2007).

2. See Batabyal (1998), Batabyal and Beladi (2002a), and Lee and Batabyal (2002) for a more detailed discussion of this point.

welfare. Therefore, in this chapter, we shed light on this general issue. Further, within the context of this general issue, we answer three specific questions for any arbitrary time period t. First, we determine the expected and the actual levels of pollution in the DC's polluting sector. Second, we compute the mean social loss arising in part from the uncertainty about the relative weight that the EA places on public versus its own welfare. Finally, we solve for the optimal value of the parameter which measures the relative weight the EA places on public versus its own welfare.

Why is it important to analyze the general issue stated at the beginning of the previous paragraph? This is because the actual practices associated with environmental policymaking in many DCs suggest that this "public versus personal welfare" issue is salient. We now corroborate this claim with some discussion of actual practices of environmental policymaking in two large and important DCs, namely, China and India.

1.1. *China*

In the case of China, the work of Sinkule and Ortolano (1995) tells us clearly that conflict of interest issues abound in the implementation of environmental policy. Consider the case of Chinese environmental protection bureaus (EPBs). Sinkule and Ortolano (1995, p. 79) note that the increased influence of EPBs may well "be offset by potential conflicts of interest that limit the EPB's ability to regulate." Citing Qu Geping, a former administrator of the National Environmental Protection Agency (NEPA), Sinkule and Ortolano (1995, p. 178) emphasize "the importance of preventing corruption and misuse of fees collected by environmental protection units . . . ." These authors also point out that very few EPBs are actually interested in seeing pollution discharge fees being set equal to the cost of treating wastewater. This is because if these fees are set as they ought to be set then factories will "respond by building more wastewater treatment plants, and then the EPBs [will] lose fees as a source of revenue" (Sinkule and Ortolano, 1995, p. 180).

More recently, Michael Palmer (2000) has commented on environmental policymaking in contemporary China. According to the revised Criminal Law, which came into effect on 1 October 1997, "officials responsible for supervising and managing the protection of the environment may be liable for criminal punishment ... for deviant acts committed in the course of duty" (Palmer, 2000, p. 73). In addition, "[a]rticles 187 and 188 stipulate liability for maladministration and abuse of power in relation, *inter alia*, to environmentally polluting conduct" (Palmer, 2000, p. 77).

1.2. *India*

In India, environmental policymaking in general and the enforcement of environmental regulations in particular leave a lot of room for improvement. For instance, in his detailed analysis of environmental policies and regulations in India, Dwivedi (1997, p. 99) points out that "the administrative machinery set up to implement the [environmental] legislation interprets its own duties from time to time, and such interpretations often do not conform to the ... intent and purpose of the law." In addition, although several meaningful environmental laws exist on the books, the fact of the matter is that environmental degradation continues to quicken. This is because government "bureaucrats and industry managers have a basically mistrustful relationship" (Dwivedi, 1997, p. 215). To make matters worse, this mistrustful relationship has "resulted in the inability of government regulatory agencies to communicate candidly and freely with industry, and in industry's reluctance to seek joint industry–government solutions to industrial pollution-control problems" (Dwivedi, 1997, p. 215).

At the level of environmental inspections, there is a considerable amount of corruption to contend with. We learn that "environmental inspectors succumb to bribes partly because they are poorly paid, partly because of the political culture prevailing in the nation and partly because the punishment is not severe enough to deter them or the polluter. In other words, both inspectors and polluters have an incentive to cheat" (Dwivedi, 1997, p. 127). This saturnine state of affairs has the unfortunate effect of making bribery rampant in society. Indeed, bribery "is the best known means of evading law enforcement, and when it is subtly employed it can be a useful delaying tactic for the polluter" (Dwivedi, 1997, p. 204). Because of the reasons given in this and the preceding paragraph, it is not unreasonable to contend, as Dwivedi and Vajpeyi (1995, p. 65) have, that environmental regulators in India suffer "from the lack of political support and public credibility. ..."[3]

1.3. *Discussion*

This discussion of environmental policymaking in both China and India and our intuition together tell us that there is really no reason to believe that an EA in a DC will only be interested in the welfare of the public. In fact, given the discussion in Sections 1.1 and 1.2, what is more likely is that as far as the implementation of environmental regulations is concerned, an EA will be interested in both public and its own welfare. However, what relative weight an arbitrary EA will place on public versus private welfare is typically not something that is known with certainty. Therefore, in this chapter, we suppose that this relative weight is a random variable. We now proceed to study the nature of the interaction between an EA and the polluting sector in a DC when there is uncertainty about the relative weight that this EA places on *public* versus its *own* welfare.

The rest of this chapter is organized as follows. Section 2.1 delineates our stylized model of the interaction between an EA of the sort described in the previous paragraph and the polluting sector in an arbitrary DC and for an arbitrary time period t. Section 2.2 ascertains the expected pollution level and the actual pollution level in the DC's polluting sector. Section 2.3 computes the mean social loss arising in part from the uncertainty about the relative weight

---

3. For more on the importance of credibility in environmental policy in DCs, see Batabyal (1998), Batabyal and Beladi (2002a, 2002b), and Lee and Batabyal (2002).

that our EA places on public versus its own welfare. Section 2.4 calculates the optimal value of the parameter — portraying the relative weight the EA places on public versus its own welfare — that minimizes the expected social loss computed in Section 2.3. Finally, Section 3 concludes and offers suggestions for future research on the subject of this chapter.

2. Public Versus Personal Welfare

2.1. *Preliminaries*

As in Batabyal and Beladi (2002b), consider a trading DC whose economy is dualistic. One sector is the traditional sector in which there is no pollution. The second sector is the modern or the industrial sector in which production causes pollution. In the remainder of this chapter, our attention will be on this polluting sector. Further, the subscript t on a variable will refer to the time period under consideration. Because we want to work with pollution in a time period, x_t, directly as the EA's control variable, as in Chapter 12, we shall assume that the functional relationship between the production of the polluting sector in time period t, q_t, and the pollution generated in this same time period, x_t, is strictly monotonic. As in Chapter 12, we would, once again, like to work with a loss function. Therefore, it will be helpful to think of the EA as an entity that sets pollution levels (the bad) directly.

To reiterate, x_t is the period t pollution level that is set by the EA in our DC. Let x_t^e denote the polluting sector's period $t-1$ expectation of what pollution will be in period t. Assuming that all agents in the polluting sector of our DC have rational expectations, we get $x_t^e \equiv E_{t-1}x_t$, where $E[\cdot]$ is the expectation operator. There will generally be some discrepancy between the EA's targeted output level of the polluting good[4] and the actual output level. To account for this, let $w > 0$ denote the positive wedge between these two

---

4. As discussed in the previous paragraph, x_t and q_t are functionally related in a specific way. Consequently, targeting pollution directly has the effect of targeting output indirectly.

output levels. In addition, the production of the polluting good may be subject to output supply shocks. To model this possibility, we let z_t be a conditional mean zero, independently and identically distributed (i.i.d.) output supply shock. Finally, we suppose that the EA's preferences over pollution and the production of the polluting good can be described by a loss function with the following form[5]

$$L_t = (x_t - x_t^e - z_t - w)^2 + \chi x_t^2 + 2\lambda_t \delta x_t. \qquad (1)$$

The loss function in Equation (1) is clearly the sum of three terms. The first term, $(x_t - x_t^e - z_t - w)^2$, represents the EA's concern for the output of the polluting good. The second term, χx_t^2, represents the EA's concern for pollution. Finally, the third term, $2\lambda_t \delta x_t$, represents the EA's concern for its own welfare. The reader may find it useful to think of this third term, $2\lambda_t \delta x_t$, as a monetary payment to the EA that is reduced when pollution increases. The parameter χ in the second term, χx_t^2 measures the cost of pollution relative to that of suboptimal output. In the third term, $2\lambda_t \delta x_t$, $\lambda_t > 0$ is a random variable that captures the relative weight our EA places on public versus its own welfare (monetary payment). To keep the subsequent mathematics straightforward, we assume that $E_{t-1}[\lambda_t] = 1$ and that the variance $Var_{t-1}[\lambda_t] = \sigma_\lambda^2$. In summary, Equation (1) tells us that our EA wishes to minimize the weighted sum of three terms that reflect its concern for the output of the polluting good, pollution itself, and its own monetary payment or welfare. The outstanding task before us now is to determine the expected and the actual pollution levels in the polluting sector of the DC under consideration.

2.2. Expected and Actual Pollution

The reader should think of the interaction between the EA and the polluting sector in our DC as a one-shot game. We now want to determine the equilibrium of this game. In symbols, we want to

---

5. This kind of loss function has been used in the monetary economics literature by Barro and Gordon (1983), Backus and Driffil (1985), and others. For a good account of dynamic consistency issues in monetary economics, see Obstfeld and Rogoff (1996, pp. 634–658).

determine the optimal values of x_t^e and x_t. We begin by solving the EA's optimization problem. This EA solves

$$\min_{\{x_t\}} L_t = (x_t - x_t^e - z_t - w)^2 + \chi x_t^2 + 2\lambda_t \delta x_t. \qquad (2)$$

The first-order necessary condition for an optimum to this problem is

$$x_t - x_t^e - z_t - w + \chi x_t + \lambda_t \delta = 0. \qquad (3)$$

Now, taking time period $t-1$ expectations, setting $E_{t-1}[x_t] = x_t^e$, and then simplifying the resulting expressions, we get

$$x_t^e = \frac{w - E_{t-1}[\lambda_t]\delta}{\chi} = \frac{w - \delta}{\chi}. \qquad (4)$$

Equation (4) gives us the equilibrium expected level of pollution. Now, to obtain the equilibrium actual level of pollution, x_t, let us substitute the above value of x_t^e from Equation (4) into Equation (3) and then solve the resulting expression for x_t, keeping in mind that $w - \delta = \chi x_t^e$. This gives us

$$x_t = x_t^e - \frac{(\lambda_t - E_{t-1}[\lambda_t])\delta}{1+\chi} + \frac{z_t}{1+\chi} = x_t^e - \frac{(\lambda_t - 1)\delta}{1+\chi} + \frac{z_t}{1+\chi}. \qquad (5)$$

Recall that the random variable $\lambda > 0$ captures the weight that our EA places on public welfare versus its own welfare. Given this interpretation, Equations (4) and (5) together describe the equilibrium of the one-shot game that we are analyzing. In addition, these two equations also tell us that when there is uncertainty about an EA's intentions as far as public versus private welfare is concerned, the expected or mean amount of pollution is the same as when λ is known to equal unity. This conclusion follows because $E_{t-1}[\lambda_t] = 1$. However, the reader should note that the *ex post* uncertainty about the type of EA that our polluting sector is confronted with creates additional variability in the actual amount of pollution that arises. We now proceed to compute the mean social loss arising in our DC in part from the uncertainty about the relative weight that our EA places on public versus its own welfare.

2.3. Mean Social Loss

To calculate the mean social loss in a straightforward manner, we shall make two assumptions. In particular, we suppose that the relevant social loss function is of the form

$$L_t = (x_t - x_t^e - z_t - w)^2 + \chi x_t^2, \qquad (6)$$

and that the covariance between the random variables λ and z is zero or $Cov(\lambda, z) = 0$. Now, making the appropriate substitutions from Equations (4) and (5) into Equation (6), we get

$$E_{\{t-1\}}L_t = E_{t-1}\left[\left\{-\frac{(\lambda_t - 1)\delta}{1+\chi} + \frac{z_t}{1+\chi} - z_t - w\right\}^2 \right.$$
$$\left. + \chi\left\{\frac{w-\delta}{\chi} - \frac{(\lambda_t-1)\delta}{1+\chi} + \frac{z_t}{1+\chi}\right\}^2\right]. \qquad (7)$$

After several steps of algebra, the right-hand side (RHS) of Equation (7) can be simplified to

$$E_{\{t-1\}}L_t = w^2 + \frac{(2-\delta)^2}{\chi} + \frac{\sigma_\lambda^2 \delta^2}{1+\chi} + \frac{\chi \sigma_z^2}{1+\chi}, \qquad (8)$$

where σ_λ^2 and σ_z^2 are the variances of the random variables λ and z, respectively. Inspection of Equation (8) leads to three salient conclusions. First, as the parameter χ which measures the cost of pollution relative to that of suboptimal output increases, the expected loss to society decreases. Second, as the uncertainty associated with the output supply shock (σ_z^2) goes up, the mean loss to society also goes up. Finally, when the uncertainty associated with the EA's weight over public versus its own welfare (σ_λ^2) increases, once again, the expected loss to society also increases. This last result clearly tells us that from the standpoint of environmental policymaking, DCs need to ensure, to the extent possible, that individuals who are placed in positions of authority are in fact public spirited in the discharge of their official duties. The final task before us now is the calculation of the optimal value of the parameter δ, which measures the relative weight the EA places on public versus its own welfare (also see Equation (1)).

2.4. The Optimal Value of the Relative Weight Parameter

Inspecting Equation (8) it is clear that if there is no uncertainty about the relative weight the EA places on public versus its own welfare, i.e., if $\sigma_\lambda^2 = 0$, then the expected social loss is minimized by choosing $\delta = w$. In other words, in this case of certainty about the EA's type, the relative weight parameter δ is chosen so that it is equal to the positive wedge between the targeted output level of the polluting good and the actual output level of this same good.

However, when $\sigma_\lambda^2 \neq 0$, and, hence, λ is unpredictable, the previous paragraph's solution is not optimal, and we have to contend with the fact that there is a tradeoff between reducing mean pollution by choosing a positive δ and raising the variance of pollution because the EA's preferences are stochastic. Now, to determine the optimal δ, we solve

$$\min_{\{\delta\}} \left[E_{t-1} L_t = w^2 + \frac{(w-\delta)^2}{\chi} + \frac{\sigma_\lambda^2 \delta^2}{1+\chi} + \frac{\chi \sigma_z^2}{1+\chi} \right]. \quad (9)$$

Differentiating Equation (9) with respect to δ and then setting the resulting expression equal to zero gives us the first-order necessary condition for an optimum to this problem. Algebraically manipulating this first-order condition gives us an expression for the optimal value of δ and that expression is

$$\delta = \frac{(1+\chi)w}{1+\chi(1+\sigma_\lambda^2)}. \quad (10)$$

Equation (10) tells us that when there is additional uncertainty about the EA's relative weight λ, i.e., when $\sigma_\lambda^2 \neq 0$, the optimal value of * is less than w, the positive wedge between the targeted output level of the polluting good and the actual output level of this same good. Consistent with the discussion in the first paragraph of this section, the reader can inspect Equation (10) and thereby easily verify that when $\sigma_\lambda^2 = 0$, the optimal value of δ equals the positive wedge w. This completes our discussion of the computation of the optimal value of the parameter δ.

3. Conclusions

In this chapter, we shed light on a hitherto unstudied question about the nature of the interaction between an EA and the polluting sector in a DC when there is uncertainty about the relative weight that this EA places on public versus its own welfare. First, in Section 2.2 we determined the mean and the actual pollution levels in the DC's polluting sector. Next, in Section 2.3 we computed the mean social loss arising in part from the uncertainty about the relative weight that our EA places on public versus its own welfare. Finally, in Section 2.4 we calculated the optimal value of the parameter — portraying the relative weight the EA places on public versus its own welfare — that minimizes the expected social loss computed in Section 2.3.

Recently, in the context of India, Dwivedi (1997) has noted that environmental policymaking can be improved by, *inter alia*, increasing the public awareness of environmental problems and by taking steps to mitigate the venality of officials responsible for environmental management. In addition to having other benefits, these sorts of actions are also likely to diminish uncertainty about an EA's type. The analysis in this chapter tells us that as far as the reduction of expected social losses is concerned, taking the above sorts of actions would clearly be a good thing.

The analysis in this chapter can be extended in a number of different directions. In what follows, we propose two possible extensions. First, one can generalize the analysis conducted here by modeling and analyzing the interaction between an EA and the polluting sector of a DC as a repeated game. Second, with regard to the use of loss functions, it would be useful to study the nature of the interaction between an EA and a DC's polluting sector when the EA's focus is not on the minimization of social losses but instead on the maximization of the net benefit from the implementation of sound environmental policy. Studies of the conduct of environmental policy in DCs which incorporate these aspects of the problem into the analysis will provide richer accounts of environmental policymaking in DCs and this is a subject of considerable contemporary significance.

References

Atkinson, G., Dubourg, R., Hamilton, K., Munasinghe, M., Pearce, D. and Young, C. (1997). *Measuring Sustainable Development*. Cheltenham, UK: Edward Elgar.

Backus, D. and Driffil, J. (1985). Rational Expectations and Policy Credibility Following a Change in Regime. *Review of Economic Studies* 52:211–221.

Barro, R.J. and Gordon, D.B. (1983). Rules, Discretion, and Reputation in a Model of Monetary Policy. *Journal of Monetary Economics* 12:101–121.

Batabyal, A.A. (1998). Environmental Policy in Developing Countries: A Dynamic Analysis. *Review of Development Economics* 2:293–304.

Batabyal, A.A. and Beladi, H. (2002a). A Dynamic Analysis of Protection and Environmental Policy in a Small Trading Developing Country. *European Journal of Operational Research* 143:197–209.

Batabyal, A.A. and Beladi, H. (2002b). Aspects of the Theory of Environmental Policy in Developing Countries. *Discrete Dynamics in Nature and Society* 7:53–58.

Brundtland, G.H. (1987). *Our Common Future*. Oxford, UK: Oxford University Press.

Dwivedi, O.P. (1997). *India's Environmental Policies, Programmes, and Stewardship*. New York: St. Martin's Press.

Dwivedi, O.P. and Khator, R. (1995). India's Environmental Policy, Programs, and Politics. In O.P. Dwivedi and K. Vajpeyi (eds.), *Environmental Policies in the Third World*. Westport, CT: Greenwood Press.

Dwivedi, O.P. and Vajpeyi, K. (eds.) (1995). *Environmental Policies in the Third World*. Westport, CT: Greenwood Press.

Farmer, M.C. and Randall, A. (1997). Policies for Sustainability: Lessons from an Overlapping Generations Model. *Land Economics* 73:608–622.

Heal, G. (1998). *Valuing the Future: Economic Theory and Sustainability*. New York: Columbia University Press.

Jan, G.P. (1995). Environmental Protection in China. In O.P. Dwivedi and K. Vajpeyi (eds.), *Environmental Policies in the Third World*. Westport, CT: Greenwood Press.

Lee, D.M. and Batabyal, A.A. (2002). Dynamic Environmental Policy in Developing Countries with a Dual Economy. *International Review of Economics and Finance* 11:191–206.

Munasinghe, M. (2007). *Sustainable Development in Practice*. Cheltenham, UK: Edward Elgar.

Obstfeld, M. and Rogoff, K. (1996). *Foundations of International Macroeconomics*. Cambridge, MA: MIT Press.

Palmer, M. (2000). Environmental Regulation in the People's Republic of China: The Face of Domestic Law. *In* R.L. Edmonds (ed.), *Managing the Chinese Environment*. Oxford, UK: Oxford University Press.

Pezzey, J.C.V. (1997). Sustainability Constraints versus "Optimality" versus Intertemporal Concern, and Axioms versus Data. *Land Economics* 73: 448–466.

Sinkule, B.J. and Ortolano, L. (1995). *Implementing Environmental Policy in China*. Westport, CT: Praeger.

Stern, N. (2007). *The Economics of Climate Change*. Cambridge, UK: Cambridge University Press.

Stoett, P.J. (1995). Environmental Problems, Policies, and Prospects in Africa: A Continental Overview. *In* O.P. Dwivedi and K. Vajpeyi (eds.), *Environmental Policies in the Third World*. Westport, CT: Greenwood Press.

Index

ad valorem tarif, 17, 18, 20, 21, 103, 106–109, 111–113, 116–118, 120, 123, 127–132, 134, 135, 137–141

balanced trade, 149, 160, 168, 171, 174, 181, 196, 197, 207
boundary condition, 109, 111, 116, 129, 152, 156, 174, 177, 200, 204
Brundtland report, 2, 227

cartel, 126, 127
cash crop, 7–9, 57, 59–61, 63, 64
cleared parcel of forest land (CPFL), 5–11, 39–54, 58, 59, 67–72, 75, 76
commitment, 23, 24, 26, 28–30, 32, 145, 150–162, 166, 171, 172, 175–183, 190, 197, 198, 201–210, 214, 221–224
competitive seller, 17–20, 103, 105, 106, 109, 112, 113, 120, 131, 132, 137–139
conditional probability, 48, 49, 51, 52, 54
conservation, 2, 4, 17–21, 103–105, 112, 113, 117–120, 123, 124, 132, 133, 135, 139–142
credible, 18, 19, 24, 27, 28, 31, 32, 118, 119, 150, 161–163, 171, 182, 183, 197, 209, 210, 215, 220
crisis, 14–17, 87, 89–100

deficit, 30–32, 149, 160, 168, 196, 197, 207, 213, 215, 218, 219, 224, 228
developing country (DC), 4, 5, 12–15, 17, 21–33, 39, 40, 53, 54, 57, 58, 67, 68, 79–81, 84, 87–90, 94, 95, 98–100, 104, 106, 124, 145, 150, 165, 189, 213, 227–229, 231–235, 237
discretion, 30, 32, 223
distortion, 26–30, 154, 159, 165, 167, 174, 175, 178, 182, 184, 185, 191, 193, 198, 200, 201, 205, 210
dual economy, 165, 166
dynamic, 1–4, 6, 7, 9, 22, 23, 25, 27, 28, 30, 34, 39, 41, 42, 44, 53, 58, 59, 67, 69, 107, 109, 115, 117, 126, 128, 137, 145, 146, 149, 153, 154, 158, 160, 165–167, 185, 189–193, 196, 197, 207, 209, 211, 213–215, 220, 233
dynamic consistency, 109, 220, 233
dynamic inconsistency, 23, 117, 158, 213, 215
dynamic programming problem, 160, 207

ecological-economic system, 2–4, 6
employment, 22–24, 27, 30, 145–147, 149, 154, 158, 162, 165, 166, 168, 170, 176, 179, 180, 183, 185, 189, 190, 192, 193, 196, 202, 206, 209, 210, 213, 214, 228
endangered resource, 19, 119
environmental policy, 5, 21–23, 25–32, 145, 146, 149, 151, 155, 156, 158, 159, 162, 163, 165–167, 169, 171, 177–179, 182–185, 189–194, 197, 200–203, 205, 206, 209–211, 213–215, 219–225, 228, 229, 231, 237
environmental protection, 27, 30–32, 145–147, 150, 165, 166, 171, 183–185, 190, 197, 213–215, 220, 223–225, 228, 229
environmental resource base, 1, 3
Euler equation, 31, 217

Index

exhaustible resource, 4, 19, 70, 124, 125, 132
expenditure function, 148, 169, 195
export subsidy, 25–27, 167, 169, 173, 175, 178–180, 182

fallow period, 5–8, 39–41, 43–54, 58, 59, 61, 62, 64, 68
fallow state, 11, 43, 48, 49, 51, 52, 71, 72, 75, 77
fertilizer, 9–12, 67, 69–77
flood, 12–14, 79–85
food crop, 8, 9, 57, 59–64

game, 17–20, 32, 33, 103, 105–109, 113–115, 118–120, 123, 125, 126, 128, 129, 134, 141, 221, 233, 234, 237
groundwater, 81, 87

Hamiltonian, 111, 132, 139, 151, 173, 199
harvest, 5, 9, 17–21, 40, 42, 43, 58, 68–70, 106–120, 127, 128, 132–136, 138–141
high quality land, 8, 9

infant industry, 25, 28, 168, 193
infinitesimal commitment, 26, 29, 159, 180, 206
initial condition, 17, 94, 98, 99, 116, 151, 173, 199
international environmental agreement (IEA), 185
international trade, 19, 103–105, 119, 123–125, 127, 142
investment, 172, 174, 184, 198
irreversible, 15, 87, 89

jump state, 109, 116, 129, 137, 151, 172, 198

land quality accumulation, 7, 8, 57, 59, 60, 62, 64
land use, 7, 57–59, 63, 64

lax management regime, 16, 92, 93, 97, 99
likelihood function, 11, 12, 67, 69, 75, 76
limited commitment, 24, 29, 154–158, 176, 179, 202–205
logistic growth function, 128, 133
long run average cost, 11, 12, 72–76
loss function, 219, 220, 223, 225, 232, 233, 235, 237
low quality land, 8, 9

M/M/1 queuing model, 13, 81, 82
Markov chain, 16, 42, 44, 90–93, 98, 99
Markov perfect equilibrium, 23, 24, 26, 29, 155, 159, 161, 176, 179, 180, 203–206
Markovian property, 82
migration, 22, 24, 26, 147–149, 151, 152, 154, 158, 162, 163, 168–170, 172–176, 183, 193, 195, 196, 198, 199, 202, 206, 209, 210
monopolistic seller, 19, 20, 120, 123, 127, 129, 131, 132, 134, 135, 139, 141
monopsonistic buyer, 19, 120, 123, 131, 138, 141

natural capital stock, 1, 2, 4
natural resource management, 3
net social benefit, 12, 79, 80, 83
non-standard control problem, 116, 137, 151

open loop tariff, 17, 18, 103, 108, 109, 113, 118, 120, 128, 129, 134
optimization, 6, 14, 31, 33, 50, 80, 83–85, 158, 217, 234

perfect commitment, 151–154, 157, 172, 175, 176, 178, 202
personal welfare, 227, 229, 232
Poisson process, 13, 81
pollution, 22–34, 145–147, 149, 151–155, 157–159, 161–163,

165–168, 170–176, 178–180, 182–184, 189, 191–194, 196–202, 205–210, 213–216, 218–224, 227–237
pollution tax, 22–24, 27–31, 147, 151–155, 157, 161, 163, 165, 168, 170–176, 178–180, 182–184, 189, 191, 193, 194, 196, 198–202, 205–210, 213–216, 218, 219, 224, 228
public welfare, 32, 33, 234

queuing theory, 13, 79–81, 84

random variable, 6, 32, 33, 42, 43, 52, 220, 221, 231, 233–235
rangeland, 2, 4, 87–90, 94, 98, 99
rational expectations, 28, 115, 116, 137, 147, 163, 169, 184, 194, 203, 210, 219, 221, 232
renewable resource, 2, 4, 12, 14–21, 87–91, 94, 98–100, 103–106, 108, 112, 113, 115, 117–120, 123–127, 131, 132, 138, 140–142
renewal theory, 73
renewal-reward theorem, 73
revenue function, 26, 27, 148, 152, 153, 157, 169, 175, 179, 191, 194, 195

safe drinking water (SDW), 4, 12, 79–85
safe minimum standard, 90
second best, 22, 24, 29, 30, 147, 162, 168, 174, 191, 193, 201, 205, 209
self-financing constraint, 30–32, 213–216, 219, 224, 228
semi-Markov processes, 41, 42, 44, 53
shifting cultivation, 4, 5, 7, 39, 40, 57, 59, 64, 95
slash and burn agriculture, 4, 5, 8, 9, 39–41, 53, 57–60, 63, 64, 67
soil fertility, 10, 11, 40, 42, 57, 67–77

Stackelberg differential game, 17, 19, 20, 103, 105, 113, 119, 123, 125, 126
stationary probability, 14, 44, 45, 49, 82, 83, 85
stochastic, 1–4, 6, 7, 9, 10, 12, 14, 34, 35, 39, 41, 42, 44, 53, 58, 59, 67, 69, 71, 76, 77, 79, 80, 84, 85, 89, 236
stochastic process, 10, 42, 91
stock dependent cost function, 18–21, 103, 105, 107, 113, 115, 116, 118, 119, 123, 125, 128, 134–136, 139, 140
stock independent cost function, 18, 20, 21, 103, 105, 107–109, 115, 118, 123, 125, 128, 133, 134, 138, 140
strict management regime, 16, 17, 92, 94–97, 99
surplus, 30–32, 168, 213, 215, 218, 219, 224, 228
sustainability, 2, 6, 39, 53
sustainable development, 2, 3, 87, 165, 213, 227, 228
swidden agriculture, 4–6, 10, 12, 39–41, 43, 47, 50, 53, 54, 57, 67, 68, 77

transition, 6, 42–44, 48, 49, 52, 54, 91, 93, 98, 99
transversality condition, 156, 178, 204

uncertainty, 33, 34, 223, 224, 227–229, 231, 234–237
unconditional probability, 44, 50, 54
unit tariff, 106, 107, 109–113, 116, 127, 129, 133, 135
utility, 19, 21, 31, 106, 109, 110, 112, 114, 117, 120, 125, 127, 129, 133–135, 140, 216–218

Wiener process, 10, 11, 71–76